SLAP SHOT SCIENCE

SLAP

ALAIN HACHÉ

SHOT

A Curious Fan's Guide to Hockey

SCIENCE

JOHNS HOPKINS UNIVERSITY PRESS   BALTIMORE

Johns Hopkins University Press
2715 North Charles Street
Baltimore, Maryland 21218-4363
www.press.jhu.edu

Library of Congress Cataloging-in-Publication Data

Haché, Alain, 1970–
Slap shot science : a curious fan's guide to hockey / Alain Haché.
pages cm
Includes index.
ISBN 978-1-4214-1792-9 (pbk. : alk. paper) —
ISBN 1-4214-1792-8 (pbk. : alk. paper) —
ISBN 978-1-4214-1793-6 (electronic) —
ISBN 1-4214-1793-6 (electronic)
1. Physics.   2. Hockey.
3. Force and energy.   I. Title.
QC28.H234 2015
530—dc23   2015002494

A catalog record for this book is available from the British Library.

Special discounts are available for bulk purchases of this book.
For more information, please contact Special Sales at 410-516-6936 or
specialsales@press.jhu.edu.

Johns Hopkins University Press uses environmentally friendly
book materials, including recycled text paper that is composed of
at least 30 percent post-consumer waste, whenever possible.

For Sophie, Amelia, Ethan, and Venitia

CONTENTS

PREFACE

When my previous book, *The Physics of Hockey*, was published in 2002, it gave hockey fans a fresh look at this complex and unique sport. While most books about hockey deal with history and technical elements of training and coaching, *The Physics of Hockey* gave readers a different perspective by examining the sport through the lens of science and technology. I certainly had plenty to discuss. Think about it: hockey involves skating, shooting, blocking shots and making saves, body checking, a whole assortment of equipment, and a host of other actions.

Since then, I have enjoyed interacting with fans and the hockey community. I have given presentations to specialists and to the general public and have collaborated with companies to develop new and more effective hockey equipment. People raised interesting questions that were not answered in the book, which I attempted to address. Sometimes it was just to satisfy our mutual curiosity and at other times it was to offer guidance on projects, such as ones for science fairs. Over the decade that followed, the game of ice hockey kept evolving. New rules, new equipment, and new strategies were introduced, while sports scientists progressed in their understanding of the game. They found better ways to evaluate and train athletes and to design better skates, sticks, and protective equipment.

I took note of these questions and new developments and wrote essays and articles related to them in preparation for a new book. Given these developments, *Slap Shot Science* could have easily been a sequel to *The Physics of Hockey* just by covering the changes in the game in the years since the first book was published. But it is more than just a sequel. In addition to filling the gaps of the previous book, *Slap Shot Science* is written so that it can be picked up by any hockey fan, coach, or player and be understood easily. *The Physics of Hockey* was heavy on mathematics; in a way, it was a surprising success. Aimed at the general public, *Slap Shot Science* presents the material in such a way that all lovers of hockey, whether they are scientifically inclined or not, can easily understand it and apply it to real-life situations.

The book deals with a broad range of hockey subjects and contains some original material and ideas that have never been presented elsewhere. Yet, at the same time, it is not meant to be encyclopedic. There's already plenty available in print and online to attempt such a feat. One can find excellent videos online with hockey tips, hockey science, coaching drills, and so on. I would in fact encourage readers to consult them to find examples related to the discussions in this book. Among my favorites are slow-motion videos of slap shots. Another example: readers can see online videos of Jonathan Quick's famous "miracle save"—and physics puzzle—that I analyze in depth in chapter 5. The book doesn't try to summarize the knowledge that is already out there but tries to give a different point of view.

Those of you who feel science and math are outside your zone of comfort will be pleasantly surprised about how much you already know and understand. Having taught physics at university for 17 years, I know that mathematics is a powerful tool to make predictions, but I also know that we can understand much of the world around us without it. Many times we simply have a good feel for what is going on. A good example is the ability people have to accurately estimate the trajectory of projectiles like baseballs, tennis balls, and hockey pucks. When a puck is shot at the net from some distance away, it follows an upside-down U trajectory that can only be described with physics equations (and lots of them, to account for air drag and lift, among other things). To tackle this problem, we need not only Newton's laws of mechanics but also computer programs. Yet, it is remarkable that a goaltender or a player can predict fairly accurately where the puck will end up just moments after it has been shot. Think of the outfielder in baseball who runs to the exact location where the ball will fall. This skill, acquired by trial and error, is applicable to a wide range of projectile speeds and launch angles, much like the power of physics equations. The "feel" the human brain has for the mechanics is as effective as—and actually much faster than—solving Newton's equations using a computer.

Another topic regarding numbers requires an explanation: the book uses a mixture of standard and international (metric) units. Sports scientists prefer to use metric units, with good reasons, but for the reader's convenience, I use the units that are most often used in each situation. Even as a professional physicist working every day with metric units, I also have a better feel for body weight when quoted in pounds and body height when measured in feet and inches. And while I'm more familiar with speeds quoted in km/h (kilometers per hour), I can understand how some readers relate better to miles per hour—as in the 100 mph slap shot benchmark. Appendix A gives a conversion table for units used in this book, and appendix B provides a list of common and not-so-common scientific symbols and abbreviations used in the book. For the mathematically inclined reader, a detailed explanation of an equation used in the book to calculate a team's winning chances is given in appendix C.

In the book, I mostly use examples and statistics from the National Hockey League. This is not an intentional snub of the excellent hockey played in other leagues, but I have done so simply because NHL data are easily available and its players best known to me. As discussed in the levels of hockey section in chapter 7, the difference between the level of play in the NHL and other top leagues is not as large as many people think. Even hockey in amateur leagues is quite exciting to watch; in my hometown, I enjoy following the Moncton Wildcats of Quebec's Major Junior Hockey League (they have graciously given me permission to publish their pictures in the book) and especially the Université de Moncton Aigles Bleus, my alma mater's hockey team.

Similarly, I use men's hockey examples in my discussions and use the masculine gender. This is strictly for the purpose of readability and conciseness, not because I am assuming women's hockey is less interesting. On the contrary, over the past decade, women's hockey has developed into a very exciting sport to watch. For example, the Canadian Women's Hockey League has made a name for itself in recent years. When Team Canada came from behind to win the gold medal at the 2014 Winter Olympics in Sochi, Russia, it was an epic moment for female hockey in Canada. Like many other Canadians, I'm sure, I watched the end of the game over and over. So I hope this book will be enjoyed and used by female and male readers of all levels.

The book is organized as follows. The first part looks at the things that make the game of hockey: the rink, the skates, the sticks, and the basic techniques. Chapter 1 discusses the science of ice, ice making, and ice maintenance. Ice conditions, indoor and outdoor, have an impact on how the sport is played, and so do the dimensions of the ice rink. Chapter 2 follows with the science of the skate and skating. I explain important aspects of skate blade shape and sharpening, as well as the mechanics of gliding and turning, ice friction, and the biomechanics of the ankle flex. Next is shooting, discussed in chapter 3. Hockey players, even goalies, love their hockey sticks. Many consider the hockey stick to be the sport's most useful piece of equipment, each player and goalie having his own set of preferences for the shape and curvature of the blade, stick length, and shaft rigidity. I also explain shooting techniques and the effects of aerodynamics and puck spin on the puck trajectory.

The next part of the book looks at the players and what can happen to them during a game. In chapter 4, I cover aspects related to performance hockey. This is where I deal with extreme performances, whether it is the fastest hockey skaters or the hardest shooters. I also present a statistical study on how age affects hockey fitness as well as the science (and art) of identifying the future elite players. Chapter 5 deals with goaltending, a favorite topic of mine since I am an active goalie. I examine the mental aspects of the position and the recent tendency of the NHL to recruit taller goalies. I also relate a real-life anecdote explaining what happens when a beer league goalie (myself) tries to stop NHL players. Chapter 6 examines common hockey injuries and the protective equipment designed to avoid them. A discussion of the sport's well-known physicality is presented from the vantage point of science and of a hockey fan.

The book ends with an exploration of hockey from a purely statistical point of view. There is much we can do with the numbers in understanding hockey and making predictions, for example. But at the same time, there is much we cannot do with statistics, and this is often the part that is forgotten. To borrow a line from statisticians, this chapter helps the reader distinguish between the signal (meaningful numbers) and the noise (random fluctuations). I suspect hockey fans, hockey pool players, coaches, and analysts will find the material useful to properly interpret daily trends and events in the hockey world, and they may return to it periodically because of it.

I wish to thank Johns Hopkins University Press and editor Vincent Burke for allowing me to proceed with this book. I am especially grateful to Pierre P. Ferguson, a colleague and physics professor at Université de Moncton, Shippagan, for providing stimulating discussion and gathering some of the data used in this book. His keen eye during manuscript revisions was much appreciated. I acknowledge Louis Poirier for helpful discussion on the physics of ice and skating, and Jay Worobets of Nike Sports Research Labs for useful conversations on the science of hockey sticks. Jason-Andrew Mackenzie's input on athlete performance was most valuable. Thanks to Ovide Theriault for graphic designs and to Prof. Gérard J. Poitras and René Thibault from Mechanical Engineering at Université de Moncton for calculations on hockey puck aerodynamics. I am also grateful to Michele Callaghan and Maria E. denBoer for their editing work. I would also like to thank Bell Media television producer Henry Kowalski for inviting me to participate in the shooting of the hockey documentary *Scoring with Science*.

Finally, for more information or to send questions to the author, please go to www.slapshotscience.org.

FIRST PERIOD

The Things That Make the Game:
Where We Play and How We Do It

CHAPTER ONE

The Rink

The basic idea behind ice hockey—two teams trying to score goals in each other's nets—is a concept that is not exceptional in any way; it is shared by many other sports, from handball and basketball to soccer and lacrosse. But two things make hockey different from other sports: the rink and the ice. The playing surface is delimited not by lines on the ground but by boards on which glass windows are mounted (for reasons that I will explain later). You would be hard-pressed to find another team sport that is watched from the other side of walls and behind glass. The skating on ice makes hockey even more peculiar as a team sport, adding a level of difficulty, one mastered after years of practice, as well as a new level of danger, from the quickly moving players and the sharpness of skate blades. These walls and the unique dangers of skating on ice are the reasons that hockey players wear equipment that is more elaborate and protective than for other sports, more so even than that for high-impact sports like American football.

In this part, we focus on these aspects of ice hockey that make the sport unique.

The Ice

To better understand the ice that coats the rink, the ice that is carefully smoothed by the Zamboni after each period of play, I first examine the essential characteristics of ice. We start by considering why not all ice is created equal.

Clear Ice, White Ice

Ice cubes, you may have noticed, don't all look the same. Some are clear and transparent while others are cloudy and whitish to varying degrees. There are two common forms of ice, called "white ice" and "clear ice." They have the same chemical composition—they are both water molecules arranged in a crystal—but their arrangement is slightly different. White ice is made of many small ice crystals (millimeters in size) packed against each other. When a ray of light enters white ice, it tends to scatter in all directions each time it crosses the edges of these tiny crystals, and this makes the ice look milky. Clear ice, in contrast, is made of large crystals, centimeters in size, and it has fewer fractures and cracks, so it does not scatter light as much. Some clear ice is so transparent that you can make lenses with it; in fact, some resourceful campers have shaped clear ice from frozen lakes into a magnifying glass and started a fire by focusing sunlight. Clear ice is not just prettier, it is also harder and allows for faster skating. This is the kind of ice preferred by speed skaters and hockey players.

Unfortunately, clear ice needs special conditions to form, and Mother Nature does not always cooperate. In fact, water does not generally freeze into big blocks of clear ice. As the temperature drops on lakes and ponds, ice often begins to form as snowflakes fall on the water. Because snowflakes are irregularly shaped ice crystals, the ice grows around them in a disorderly fashion, making the ice look white. Over time, though, as the ice sheet grows thicker, a layer of clear ice will often appear underneath it. This is because, at the bottom, the ice has a chance to grow in a more uniform fashion from cold water instead of snowflakes. Underneath the ice sheet, the variation in temperature in the vertical direction promotes the growth of large, uniform ice crystals. In figure 1.1, a photo taken on Dow's Lake near the Rideau Canal in Ottawa, the two types of ice are easily distinguishable.

Just like outdoor clear ice, indoor clear ice requires special conditions to form. Under normal circumstances, humidity inside a hockey rink tends to create frost at the surface of the ice. Like snowflakes on a lake, frost promotes the growth of a rough layer of white ice. Skate blades make the problem worse by fracturing and damaging the ice repeatedly. This is why the Zamboni resurfaces the ice between the periods of a hockey game. The ice-resurfacing machine shaves off the layer of white ice and replaces it with hot water that freezes into clear ice. This regular operation is necessary to maintain a sheet of high-quality clear and hard ice. Yet, as time goes on, the scattering of light by the white ice from the accumulated damage will cause the

FIGURE 1.1. In this sample of lake ice, the top layer is white, as it originated from ice that grew in a disorderly way from snowflakes, while the clear section developed underneath from cold water. The hand gives an idea of scale. (Photo courtesy of Louis Poirier, National Research Council of Canada)

lines and markings to appear dimmer and fuzzier. But when the ice is new and clear, the lines appear bright blue and red. (White ice is not to be confused with the ice looking bright white, which is due to a layer of white paint put underneath the lines and markings for better contrast.)

Heat and Ice

Ice has many amazing characteristics, starting with its trademark feature, slipperiness. Equally remarkable, its liquid form, water, is one of the rare compounds that do not shrink when freezing but rather expand. This increased volume is what causes ice cubes to float in water. The same property of ice has expensive consequences in countries of higher latitudes, as it causes structural damage to roads and buildings during the seasonal cycles of freezing and thawing.

There's another surprising property of ice that is relatively less known yet easily demonstrated with a home experiment: put a pot of water on the stove, bring it to a boil, then add to it the same amount of ice by weight. (You can do this by freezing the same amount of water beforehand.) Remove the pot from the stove, mix the ice and water well, and let it all melt. Can you guess the final temperature of the mixture? Would it feel cold or hot, and would you even dare dip your finger into it?

We would expect the final temperature to be somewhere between the temperature of ice (0 °C) and the temperature of boiling water (100 °C). A reasonable guess would be the middle point, around 50 °C. It does make intuitive sense, but in reality, the final temperature would be a frigid 10 °C. Ice seems to win over boiling water by a wide margin. Why?

As it turns out, it takes a lot of heat just to melt ice. One kilogram of ice needs 334,000 Joules[1] of energy to melt, that is, to go from ice at 0 °C to water at 0 °C. This energy is called the "latent heat of fusion." Once the ice has turned into water, it takes only 4,200 Joules of energy to raise 1 kilogram of it by 1 °C. This is why it takes as much heat to melt ice as it takes to raise the temperature of the melted water to 80 °C.

Freezing, the opposite of melting, requires the opposite transfer of energy. To freeze 1 kilogram of water that is already at 0 °C, some 334,000 Joules of energy have to be removed from it. For the people in charge of making ice for hockey, this is important because 40 metric tons of water go into making a typical hockey rink. Cooling this water from room temperature down to −5 °C, the typical temperature used for hockey, requires more than 15 billion Joules of heat to be extracted and then some, because the cooling system must keep the ice below freezing for the entire hockey season. The heat extracted when freezing 40 metric tons of water would be enough to power an average household for 2 months. Once frozen, the ice temperature is easier to control because ice takes 2,100 Joules of energy per kilogram to vary its temperature by 1 °C. In this respect, ice and water have similar properties; it is the transition between the two that expends a high amount of energy.

Creating the Ice Sheet

As a boy, I used to think that the ice at my local arena was something like 10 centimeters thick (about 4 inches), about the same as

---

1. 1 Joule is the amount of energy produced (or consumed) by 1 watt of power in 1 second.

the ice of frozen rivers and ponds in the middle of winter. In reality, it is about 3 centimeters (or 1 inch) thick, and sometimes less: just enough for the cooling system to keep it frozen, yet not so thin as to be perforated by skate blades.

The process of making artificial ice starts with the cooling of the rink floor, a concrete slab through which zigzags a network of cooling pipes. As we saw in the previous section, a temperature between −5 and −10 °C is typical for hockey. A fine mist of water is applied over the whole rink with a machine, like a modified Zamboni with a water-spraying system. After several applications, the mist turns into a thin layer of ice, and a hose is used to spread more water to thicken the ice. The thin mist creates millimeter-sized ice crystals on the concrete slab. With the refrigeration system drawing heat downward, vertical ice crystal growth is favored, and large, vertically oriented ice crystals appear atop the first 1 to 2 millimeters of the ice.

The artistic part of the job now begins. White paint is applied all over the surface, either sprayed by hand or with an ice-resurfacing machine adapted for the task. The paint, a mixture of water and small white particles, freezes shortly after application. This layer gives the ice its bright white color; without it, the ice surface would take on the color of the concrete floor, a dull gray, as in the early days of hockey. The paint is then covered by another thin layer of ice, on which lines, circles, markings, and logos are laid out, usually by means of paper-like layers, or with paint applied the old-fashioned way, by hand, with brushes, rollers, and stencils.

In the final stages, the rink is flooded with hoses to add the last layers of ice. In the end, the ice is about 3 centimeters thick. Each centimeter of ice on a North American rink (or the official National Hockey League, or NHL, rink) uses 15,200 liters (4,000 gallons) of water. For the larger International Ice Hockey Federation (or IIHF) hockey rinks, it takes 17,800 liters (4,700 gallons) per centimeter of ice.

Keeping It Smooth

At the end of a period in a hard-fought hockey game, the ice is covered with deep cuts and ice shavings (called "snow") from the players' skates shaving the ice. These cause the puck to slow down and bounce. Yet, in a matter of minutes, the ice is made to shine again thanks to the ice-resurfacing machine, something hockey fans often take for granted. The ice-resurfacing machine, also called the "Zamboni" for the company by the same name, is a machine capable of bringing the ice back to its optimal playing condition in a single sweep.

The "single sweep" is the tricky part; it is actually a delicate job involving four stages. First, a thin layer of ice is shaved off with a sharp steel blade. The operator controls the blade's angle and depth (usually with a small wheel to his right) as the vehicle moves. Second, the snow is collected with rotating screws and conveyed to a snow tank, located at the front of the vehicle. Next, water is poured on the ice and then sucked back in to clear the surface of debris. Finally, a thin layer of water at a temperature of about 65 °C is spread evenly, which comes steaming underneath a pad. In all, 500 liters of water are spread over the entire rink just to make less than a millimeter of new ice.

Sliding on Thin Ice

There are few substances in Nature that offer less resistance to rubbing than ice. The rubbing surfaces inside skeletal joints are one example, but in a way the comparison is inadequate because ice is a simple chemical compound while joints are complex biological tissues. Ice is about 100 times more slippery than most surfaces we find in daily life. This is one of the most extreme features of ice and one that makes winter sports possible, including those played on snow.

What makes ice slippery—and skating possible—has puzzled scientists for centuries. What does it mean to be slippery in scientific terms? Simply put, a slippery surface is one that offers little resistance to sliding. This resistance force, or friction force, is the one we actually use to walk: with no friction between our shoes and the floor, we would be unable push ourselves forward or in any other direction.

We measure how slippery things are by calculating a friction coefficient, which is the friction force expressed as a percentage of the weight of an object (assuming the object is resting on a horizontal surface). For example, a friction coefficient of 0.2 means that the friction force represents 20 percent of a body's weight. Surfaces like concrete and wooden floors offer friction coefficients of roughly 0.5. Ice, in contrast, has a coefficient of between 0.003 and 0.05.

Friction and its effect on moving objects is a fairly straightforward phenomenon that we teach in introductory physics classes. Yet at the same time it is very complex. When two surfaces rub against each other, things happen at the microscopic level. No surface is perfectly smooth, and small bumps and pits grip each other and offer resistance to sliding. Things also happen on a smaller scale, at the molecular level. Sliding causes molecular vibrations, material deformation, displacement, and sometimes melting, to name just a few of the processes. To accurately calculate the friction force resulting from all these interactions is an impossible feat, even with the best physics models and computers. And from this complexity emerge some counterintuitive effects. For example, rough surfaces, like sheets of sandpaper, offer large friction forces, and smoothing the surfaces by polishing tends to lower friction. Yet in other instances, if you keep polishing a surface, friction will increase again because the interaction between the surfaces eventually occurs at an atomic level, where molecular bonds take over. For example, highly polished surfaces, like sheets of glass, tend to stick together more than when they are slightly rough.

In the case of a skate blade sliding on ice, there is an additional source of friction caused by ice deformation. Ice deformation causes something called "internal friction," which requires additional efforts by the skater: he has to do work (expend energy) for the blade to leave a permanent groove on the ice. This is different from what scientists call "dry friction," the natural slipperiness of ice itself.

So, before discussing the friction of ice, let's keep in mind that it is a complicated problem that depends on a number of variables, some of which may hardly be measurable. In experiments done in laboratories with a carefully controlled environment, scientists often obtain ice friction results that are different, so imagine what happens in real life, where environmental conditions change all the time. I will therefore limit the discussion to some of the general principles of ice friction and to what is most relevant to hockey.

Here are the factors that play a role in determining the friction force between a skate blade and the ice. (Note that we look at the skate in more detail in the following chapter.)

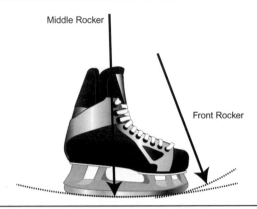

Middle Rocker

Front Rocker

FIGURE 1.2. The hockey skate blade has a rounded shape for better maneuverability. The radii of curvature of the blade are not the same everywhere, as indicated by the front rocker and the middle rocker. The radius of curvature is given by the size of the imaginary circle that matches the shape of the blade at a particular point.

■ *Blade shape*: The underside of a hockey skate blade (figure 1.2) is not flat but rounded with uneven curvature to allow good support from the ice yet quick turning and maneuverability. The blade curvature makes the area of contact with the ice relatively short, thereby putting more pressure on it. The resulting ice deformation (and internal friction), seen as the groove left in the trail, tends to slow down the skater. In speed skating, internal friction is reduced by the use of flatter, longer skate blades.

■ *Blade width*: It is possible to lower friction on a skate blade by as much as 30 percent[2] by flaring the bottom of the blade, that is, by making it wider at the bottom than at the top (like the shape of the legs of flare pants). This prevents the sides of the blade from rubbing against the ice and slowing down the skater during a sharp turn.

■ *Ice temperature*: Warmer ice is softer and tends to deform more under the skate, thereby increasing internal friction. Yet at the same time, warmer ice is more slippery (less dry friction). The two

2. See the CT Edge skate blade and P. A. Federolf, R. Mills, and B. Nigg, "Ice Friction of Flared Ice Hockey Skate Blades," *Journal of Sports Sciences* 26, no. 11 (September 2008): 1201–8.

counteracting trends—softness versus slipperiness—yield an ideal ice temperature between −5 and −10 °C[3] where the overall friction is minimized.

- *Skating speed*: At high speeds, the skater needs to push harder against the ice and the blade digs deeper into the ice. Speed tends to increase internal friction.

- *Blade material and texture*: Steel is the most commonly used metal for skate blades, but there are other materials, like nickel and titanium alloys, that have been shown to offer lower friction coefficients.

As mentioned before, the friction coefficient of a typical steel skate blade on ice ranges from 0.003 to 0.05, and sometimes more depending on ice conditions and skating speed. Multiplying this coefficient by the weight of a sliding object gives the friction force. To give an example, with a friction coefficient of 0.05, a 100 kilogram skater would experience a friction force of 50 N (the Newton is a standard unit of force), or the weight of a 5 kilogram mass.[4] As a result, to maintain constant speed, the skater needs to produce the same force forward, expending energy in the process. The power dissipated by the skater in doing so is obtained by multiplying the force by the velocity. At moderate skating speeds (say 7 meters/second, or 15 mph), the above player would dissipate 350 watts of power against the ice. But this represents only a portion of the athlete's power expenditure during a game, the rest coming from acceleration (turns and stop-and-go motion), checking, battles for the puck, and air friction.

It is possible to reduce ice friction and energy expenditure by keeping the skate blade warm. Electrically heating the skate blades, for example, with a ThermaBlade, has been shown to reduce friction by half. Oxygen consumption by the skater, a measure of physical exertion, drops by 10 to 15 percent at moderate skating speeds with heated blades. Continuously heating the blade offers a double advantage of

---

3. Jos J. de Konning, G. de Groot, and G. Jan Van Ingen Schenau, "Ice Friction during Speed Skating," *Journal of Biomechanics* 25, no. 6 (1992): 565–71.

4. In common language, the words "weight" and "mass" are used interchangeably, but in science they have different meanings: mass is a measure of inertia (the quantity of matter) and is measured in kilograms or pounds, whereas weight is the pull of gravity on a mass, a force that is measured in Newtons.

making the surface of ice warmer, thereby lowering dry friction, without making it softer and increasing internal friction.

Indoor versus Outdoor Ice

Indoor hockey is played at ice temperatures varying between 0 and –10 °C, with typical temperatures near the –5 °C mark. Outdoor hockey is at the mercy of the elements, so temperature can vary widely, although few will venture out when it dips below –15 °C. Yet even high-level hockey has been played in very cold environments. The 2003 Heritage Classic between the Montreal Canadiens and the Edmonton Oilers was played in an open stadium at –20 °C with nearly 60,000 fans cheering on.

In warmer climates, where winter temperatures are above the freezing point, the ice can be maintained within the desired range with the help of robust cooling systems. Such was the case for the 2011 NHL Winter Classic, when the Pittsburgh Penguins hosted the Washington Capitals in temperatures hovering around 10 °C. The NHL's first warm-weather outdoor game was played in 2014 at Dodger's Stadium in Los Angeles on a balmy 17 °C evening.

Outdoor hockey's enemy number one is rain, and enemy number two is snow. Rain is bad because it softens the ice and slows the puck down. When the water freezes, it makes the ice uneven and slushy; the wind sometimes adds ripples to it. Water on ice is not good for passing pucks: the water layer causes the rubber of the puck to stick to the ice and makes it stop suddenly. This also happens when players try to handle pucks when the ice is still wet from the Zamboni's resurfacing. Snow, which builds up and creates resistance for skaters and pucks alike, slows them down and makes passing tricky.

Ice, the Synthetic Way

In the early days of hockey, the sport was restricted to times and places where outdoor temperatures reached below freezing. Then came artificial ice and indoor rinks, which made hockey playable virtually anywhere at any time, at a monetary cost that varies according to climate. Cooling an ice rink in Dubai is harder than cooling one in the mountains of Switzerland.

The introduction of synthetic ice was aimed at changing all that by avoiding the cost and complexity of cooling systems. Made of hard, dense plastic panels, synthetic ice is compatible with regular metal skate blades and unaffected by outdoor or indoor conditions. If synthetic ice sounds like a high-tech product, it is not really new. Scientists and engineers have been searching for materials that mimic the gliding properties of ice since the 1950s. Early on, polymer plastics were used, but they required the application of substantial amounts of fluids to be skated on. The use of these liquids is problematic because it leads to the build-up of dirt and dust.

Over time, research improved synthetic ice to a point where some synthetic surfaces require no liquid and are used for normal skating (though the addition of lubricant always helps to reduce friction). The polymer and lubricants are designed so that the liquid is absorbed and does not retain contaminant particles. Recently engineered materials include polypropylene (known for its use in reusable plastic containers) and various grades of polyethylene (used in plastic grocery bags). Synthetic ice does have some benefits over regular ice. Those annoying deep ice grooves are not there. Also, the ready-made panels, assembled like a giant puzzle, are cheaper to maintain than an ice rink and can be used outside in the summer.

But synthetic ice has a number of disadvantages:

- It dulls the skate blade significantly faster than ice does, to the point where a skater may have to sharpen his skates after each session. Also, the polymers have less thermal conductivity than ice to channel the heat away, so the blades get really warm.
- It glides less. With more friction, the skater has to expend more effort, making playing hockey more tiring. The higher friction is part of the reason that the skate blades get noticeably hot.
- It can be time-consuming to clean by hand (there are no Zambonis for synthetic ice, yet), and once the panels are damaged, they need to be replaced.

It still has some way to go to compete with real ice, but synthetic ice may one day have the same role in hockey as synthetic turf has in football, soccer, and baseball.

The Size of the Rink

The rinks used around the world to play hockey are all similar but not identical. There are two main standards for rink sizes: (1) the

North American standard, the one used by leagues like the National Hockey League (NHL), the American Hockey League (AHL), and the National Collegiate Athletic Association (NCAA); and (2) the international, or International Ice Hockey Federation (IIHF), standard. The Olympic Games follow the international standard, with the notable exception of the 2008 Vancouver Winter Olympics, which were played on North American ice rinks.

The main difference between the two rink standards is that the IIHF rink is 16 percent wider. It may not sound like much, but if, like me, you're used to playing and watching hockey on North American rinks, the international one truly does look and feel wider. Defensemen are farther apart and use longer cross-ice passes. There is more space, and as a whole, the playing surface area is 17 percent larger than on the North American rink.

The general dimensions for each type of rink are given in table 1.1, in both metric and standard units. The quoted numbers are to be taken as typical values since there is some margin of flexibility, especially for features like the radius of curvature of corners. The distances featured in table 1.2 are the ones that are most relevant to the game of hockey. They include the end-zone lengths, the longest possible legal pass, the shortest possible icing, and the longest possible end-zone shot on net. Figure 1.3a–c illustrates the measurements of tables 1.1 and 1.2. The greatest difference between the rinks, with respect to percentage, is the space behind the net and the neutral zone length.

As every North American player who has taken part in the Winter Olympics can attest, the size of a rink affects the way the game is played. With more space between the players, the interaction between them is somewhat reduced. There is more space for the skater to accelerate. To put a number to the amount of "playing space" given to each player, we can consider the distance between a randomly positioned player and his nearest randomly positioned opponent (the one most likely to interact with him). Table 1.3 shows how rink size and the number of players on the ice affect this shortest distance of interaction. According to these results based on computer simulations in which points are randomly chosen in the rink and average distances calculated, international rinks give players between 5 and 8 percent more space to play (which translates into between 10 and 17 percent

Table 1.1. Dimensions of North American and international rinks

Item figure 1.3a		North American		International		
		Metric (m)	Imperial (ft)	Metric (m)	Imperial (ft)	Difference (%)
A	Length	61.0	200	61.0	200	0
B	Width	25.9	85	30.0	98.4	+16
C	Corner radius	8.5	28	7.75	25.4	−9
	Playing surface area	1,520 m²	16,300 ft²	1,780 m²	19,100 ft²	+17
	Contour distance (board length)	159	522	169	553	+6
	One lap distance (staying 3 ft inside board)	153	503	163	535	+6

more area to cover because area goes as the distance squared). Of course, in real-life situations, where playing strategies like "cover your man" apply, actual distances may differ. But while the distance to the nearest neighbor may be smaller in reality than in a random situation, it should still scale with rink size in the same way. In an ever-shifting game, going from point A to point B will always take longer on a bigger rink.

With more space to maneuver, it appears that skilled players have a better chance to stand out and put their talent to good use. This is why NHL coaches like to put their fastest skaters on the ice during a four-on-four overtime.

Smaller rinks, in contrast, make the game more intimate and more physical, with the possible drawback of hooking and grabbing, thus slowing the game even more. Not surprisingly, more playing space generates more goals. The NHL's four-on-four overtime produces 17 percent more goals per minute of play than the regular five-on-five format. Notice in table 1.3 that the four-on-four format increases the distance to the nearest opponent by 13 percent in both rink sizes.

Some have argued that the NHL should adopt the international-sized rink to open up play, speed up the game, and reduce injuries. Some complain that the rink has become more crowded because

Table 1.2. Dimensions of North American and international hockey rinks

Item in figure 1.3	Description	North American Metric (m)	North American Imperial (ft)	International Metric (m)	International Imperial (ft)	Difference (%)
D	Neutral zone length	15.2	50.0	17.7	58.0	+16
E	Longest possible straight line	61.8	203	63.3	208	+2
F	Longest possible shot on goal	57.8	190	58.3	191	+1
G	Goal-to-goal distance	54.3	178	53.0	174	−2
H	Longest legal pass	42.8	141	46.4	152	+8
I	Shortest icing	27.1	89	27.1	89	0
a	End-zone length	22.9	75.0	21.7	71.1	−5
b	Space between goal line and board behind the net	3.35	11.0	4.00	13.1	+19
c	Slot-to-net distance	6.10	20.0	6.00	19.7	−2
d	Faceoff circle diameter	9.14	30.0	9.00	29.5	−2
e	Longest in-zone shot on center net	23.4	76.8	23.2	76.0	−1
f	Straight shot from the blue line to the net	19.5	64.0	17.7	58.0	−9
g	Distance between faceoff dots	13.4	44.0	14.0	45.9	+4
h	Faceoff-to-goal distance	9.06	29.7	9.22	30.2	+2
aa	Crease width	2.44	8.0	na	na	na
bb	Crease circle radius	1.83	6.0	1.80	5.9	−2
cc	Crease margin	0.30	1.0	na	na	na
dd	Space behind the net	2.24	7.3	2.88	9.4	+29

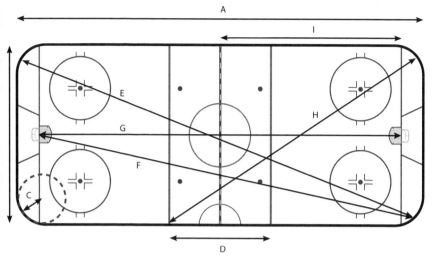

FIGURE 1.3A. Ice rink dimensions

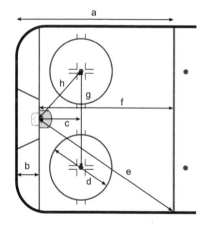

FIGURE 1.3B. End-zone dimensions

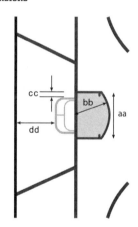

FIGURE 1.3C. Net area dimensions

athletes are becoming larger and faster. There are numbers to back up the argument. When Art Quinney and his colleagues at the University of Alberta[5] measured the average height and weight of NHL

5. H. A. Quinney, R. Dewart, A. Game, G. Snydmiller, D. Warburton, and G. Bell, "A 26 Year Physiological Description of a National Hockey League Team," *Applied Physiology, Nutrition, and Metabolism* 33, no. 4 (2008): 753–60.

Table 1.3. Playing space on North American and international rinks

	North American		International		
	Metric (m)	Imperial (ft)	Metric (m)	Imperial (ft)	Differen (%)
Average distance between two random points on the ice	22.9	75.0	24.0	78.7	+5
Average distance to the nearest of five players (e.g., opponents)	9.8	32.3	10.6	34.8	+8
Average distance to the nearest of four players (e.g., teammates)	11.1	36.5	11.9	39.1	+7
Average distance to the nearest of three players (e.g., teammates during a penalty kill)	13.0	42.7	13.9	45.5	+7

Note: I am assuming the players are randomly positioned on the ice.

players from 1979 to 2005, they found that players had become 2.3 percent taller and 8.9 percent heavier over that period. These are gains of 1 inch in height and 18 pounds in mass. In 2005, the NHL took action to open up play: the center red line was removed to allow longer passes, and referees were asked to be less tolerant of hooking and grabbing. The results were immediate and positive.

Still, some experts say that the rise in speed and size of players, unmatched by an equal increase in playing space, is the main reason for the increase in injuries and concussions. We will look at injuries more closely in chapter 6, but having a larger ice rink may not be a clear-cut solution, however, because there are effects that tend to cancel out. For example, more space makes collisions between players less frequent, but their higher speeds make them more dangerous.

The Glass

Ice hockey, along with squash and racquetball, is one of a handful of sports we watch behind glass. Ice rinks are surrounded by glass panels for the double purpose of protecting spectators from pucks and sticks and of preventing players from falling over the board when they receive a body check. The sides of the rink are mounted with panels typically extending 1.5 meters above the boards and 2.4 meters in the end zones for greater protection against pucks shot toward the net. The exceptions are the boards in front of the players' bench, which do not contain glass—and this is where players are sometimes sent overboard during a body check.

At first glance, we might think that all window panels are the same transparent material, but they are not always made of glass. Acrylic, also known as Plexiglas, is a clear plastic material used as a cheaper alternative to glass, with flexibility and resistance that make it capable of absorbing solid blows without breaking. When it breaks, it fragments into large pieces with sharp edges. One disadvantage of Plexiglas windows is that they are easily scratched. The easiest way to tell the difference between glass and Plexiglas is visually: an old glass window looks neat and clear while Plexiglas shows marks and scratches, giving it a slightly milky color. Another way to know is by touch. Glass conducts heat away better and therefore feels colder. And, when touching the window, there's yet a third way to tell the difference: push on it; if it doesn't move at all, it's glass. Meanwhile, a sheet of Plexiglas will bend by several centimeters during a hard body check, as seen in figure 1.4a.

Tempered glass is the type of glass used mainly for its strength and visual clarity. It cannot be cut or bent without shattering all at once. It's happened many times during hockey games. In 2008, Dion Phaneuf, then playing at home with the Calgary Flames, hit a glass panel behind the net with a hard shot and shattered it. Fans cheered on while the panel was being replaced. Later in the same period, he did it again, two panels over, much to the fans' delight. On the first shot, the puck hit the corner of the panel, and yet the entire window crumbled into pieces, like the one shown in figure 1.4b.

Tempered glass is made by taking ordinary molten glass and rapidly cooling it in water or air. This makes the outside glass turn solid before the inside. When the inside solidifies, it shrinks (like most materials) and, in doing so, it pulls the outer shell of the solid glass toward the center. This puts the outer layer of glass under compression stress, making it very hard, so much so that it can withstand blows by a hammer. (A neat demonstration of this is made with the so-called Prince Ruppert's drops in which hot molten glass is dropped in cold water.) But if a crack forms, even a tiny one, the stress is released and the fracture zigzags and spreads over the entire window,

FIGURE 1.4A. (*left*) Plexiglas panels used by NHL rinks are flexible and help cushion the blow from body checks, like this one by Boston Bruins' Milan Lucic on Carolina Hurricanes' Nathan Gerbe. The bending of the panel is apparent from the streaks caused by light-focusing effects. (AP Photo/Michael Dwyer) **FIGURE 1.4B**. (*right*) Tempered glass windows shatter all at once when a crack is made, like during this game between Norway and Austria at the 2014 Winter Olympics in Sochi, Russia. (AP Photo/Mark Humphrey)

turning it into a pile of pellets. This all happens in the blink of an eye. Of course, an exploding glass panel makes a body check or a slap shot all the more spectacular, but it is as much about hitting the window on a weak spot as it is applying great force.

Glass pellets are less likely to cause injury than glass shards, hence the commonly used name "safety glass." Tempered glass is used in products like furniture, glass doors, and windshields. Sometimes the glass is layered with polyvinyl to make it stand after it breaks into pellets. Glass panels must be cut into their final shape before tempering: unlike acrylic windows, they can't be cut and customized on site.

Tempered glass may be safe, but, to a hockey player slamming against it, it can feel like hitting a wall. Not only is it hard, it is also more than twice as heavy as Plexiglas, giving it some inertia. A typical sheet of glass measuring 1.5 meters wide by 2.4 meters high and 16 millimeters in thickness has a mass of 140 kilograms (300 pounds).

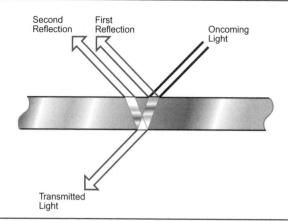

FIGURE 1.5. Light refracts (bends) and partially reflects when it travels from one medium to another. Back-and-forth reflections also occur but they are negligible compared to the first two reflections.

If a panel were to accidentally come loose, it would present a real danger. For all these reasons, acrylic products have gained in popularity recently. The NHL has made acrylic panels mandatory, with the hope that they will reduce injuries during body checks. In addition, moveable fixtures are sometimes used to allow windows to move a few extra centimeters when they are pushed. Curved panels are also used at each end of the players' bench, where some had been hurt when hitting the previously used vertical posts.

Even the best and clearest windows are never fully transparent. This is because all substances, even clear materials like glass and acrylic, reflect a bit of light (figure 1.5). Each time light goes from one type of material to another, a partial reflection occurs. Each air-glass interface of the window reflects 4 percent of light, giving a total of 8 percent bouncing back, enough so that we can see ourselves in a window when it is dark outside, but not enough that we can do so when it is bright.

The amount of light reflected off a window depends on the density of the glass, the viewing angle, and the polarization of light. When traveling at high angles, in a direction nearly parallel to the window, more light tends to reflect off it. An example is shown in figure 1.6A, where players located along the board are less visible. Spectators sitting in the front rows will notice this effect more than

FIGURE 1.6A. (*left*) The amount of light reflected off windows increases with the viewing angle, making it harder to see the end zone. (Photo courtesy of Tran Vinh Son)
FIGURE 1.6B. (*right*) At high viewing angles, the glass acts like a mirror.

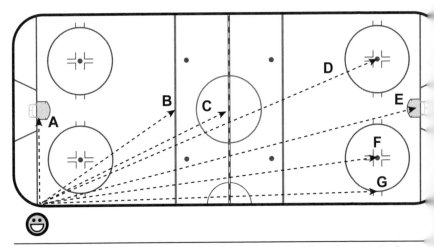

FIGURE 1.7. A viewer sitting near the boards cannot see all of the ice because of reflective glare. See table 1.4 for the percentage of light and the angle of view for fans sitting at the goal line and looking at various places on the rink.

those sitting elsewhere. From inside the rink, hockey players can use the windows as mirrors, as figure 1.6B shows.

Both acrylic and glass tend to reflect the same amount of light. The amount of light seen by a viewer sitting in the end zone and looking at various points on the ice (figure 1.7) is presented in table 1.4. For

Table 1.4. Amount of light seen by a viewer sitting at the goal line and looking at various places on the rink

Looking point	Viewing angle (degrees)	Transmitted light (%)
A	0	92
B	55	85
C	65	77
D	68	71
E	77	49
F	83	24
G	88	2

Note: See fig. 1.7 for reference.

example, a viewer sitting at the goal line and looking at the net at the opposite end of the rink receives only 50 percent of light. A common solution to reduce reflection losses is to coat glass with anti-reflection films. This would not work on rinks, however, because such coatings are effective only for a narrow range of viewing angles.

High-Altitude Hockey

Does the altitude of a hockey rink affect the way the game is played? It is a well-known fact among baseball fans that Denver, Colorado, is the place where baseballs travel the farthest. A 400-feet drive at Yankees Stadium, at sea level in New York, would travel something like 420 feet at Coors Field in the Mile-High City.[6] Each hit has a better chance to become a home run.

The main reason for this is air density: there is less atmospheric pressure at high altitude, and the thinner the air, the less drag there is on projectiles. For the athlete, this low pressure reduces aerodynamic drag while increasing the physiological demand because of lower oxygen levels.

When experts talk about training at high altitudes, they usually refer to places with elevations greater than 2,000 meters (6,500 feet)

6. A discussion of this effect can be found in R. K. Adair, *The Physics of Baseball*, 3rd ed. (New York: HarperPerennial, 2002). See also J. Brancazio, "Trajectory of a Fly Ball," *The Physics Teacher*, January 1985, 20–23.

Table 1.5. Elevation of NHL cities and corresponding air density
relative to that at sea level

Cities with NHL teams	Elevation group (m)	Air density relative to that at sea level (kg/m³)
Boston, East Rutherford, Miami, Tampa Bay, Vancouver, Washington, D.C.	0 (sea level)	1.0
Anaheim, Montreal, New York, Philadelphia, San Jose, Uniondale	0–50	0.99
Dallas, Los Angeles, Ottawa, Raleigh, St. Louis, Toronto	50–150	0.99
Buffalo, Chicago, Detroit, Nashville, St. Paul, Winnipeg	150–275	0.98
Columbus, Phoenix, Pittsburgh	275–400	0.97
Edmonton	668	0.93
Calgary	1,048	0.90
Denver	1,609	0.85

Note: For reference, the highest point on land, the peak of Mount Everest, has an elevation of 8,848 m and a relative air density of 0.38 m³.

above sea level; some even put the threshold at 2,400 meters (8,000 feet). Denver is located at an altitude of 1,600 meters (5,200 feet), so it does not quite make the cut for that definition. Still, when the Colorado Avalanche practice at home, they experience 15 percent less air drag than players do at sea level. In table 1.5, all cities with NHL teams are clustered into eight groups according to their altitude, and their corresponding air densities are given. At sea level, dry air has a density of 1.22 kilograms per cubic meter (surprisingly, humid air is a bit lighter). In Denver, air density is 1.05 kilograms per cubic meter, or 15 percent less than in sea-level cities like Tampa and Vancouver.

Considering that a portion of the energy expenditure by an athlete during skating is due to air drag, this effect is not quite negligible. Moreover, thinner air forces the body to make physiological adjustments, such as the production of more red blood cells to carry oxygen more efficiently. Athletes who regularly train at high altitudes have a slight edge over their lesser-acclimated opponents. For the visiting team that is not acclimated, the benefit of reduced air drag may be voided by the reduced oxygen, but the home team benefits from both the reduced air drag and acclimation.

Air pressure also plays a role in how the puck moves. Aerodynamic drag forces, including aerodynamic lift, vary proportionally with air density. So a puck in Denver experiences 15 percent less air drag and air lift than it does in Boston. Yet, plugging a 15 percent air drag difference into physics models does not lead to puck speed and trajectory as different as one would imagine. Part of the reason is that the dominant factor is the initial puck velocity, which is mostly determined by the effort of the player and not by air density.

There is another consequence of high altitude: the pull of gravity is weaker. As one moves away from the center of the Earth, the pull of gravity drops in accordance with Newton's law of gravitation. But because the Earth is so large, with a radius of 6,400 kilometers, one has to travel quite far before a difference is felt. So in Denver, gravity is just one-twentieth of 1 percent lower than at sea level. This is much too small to make hockey players feel any lighter or to have any impact on their performance.

Having looked at the rink and the ice where hockey is played, let's now turn our attention to the players themselves and how they are able to move and play so well on the ice.

CHAPTER TWO

The Skate

In the first chapter we looked at one of the main differences between hockey and other major sports: the fact that it takes place on ice and within walls. But there's another key difference with ice hockey: all the players are wearing skates. Some hockey fans take this for granted, but it is no small detail, certainly not to those who have never skated or to those who tried it and saw how tricky it is. I have been told by friends from countries where hockey is not played how puzzled they were at seeing hockey players skate with ease backward and forward.

Skating, like so many other things, is best learned as a child. Fear of going too fast and falling is a real barrier to learning how to skate for many adults. But, no matter the age at which we start learning, skating takes years to truly master.

Stepping on the ice for the first time, you soon realize that you will have to put away the idea that skating is like walking. Walking on the ice with shoes is easy if they have the right kind of tread, but walking with skates, you either move at a snail's pace or not at all, and there's the constant danger of falling on your back or your knees. Once the skater has learned to move around, pushing his skates side to side, there is the additional challenge of learning how to stop, which is something many people find even harder.

Children don't have to take lessons on how to walk and run, they figure it out on their own, and the same is mostly true for skating. Of course, we do tell beginners to hold on to chairs at first and, when they fall, to get back up on one foot at a time. We teach them how to bend their knees and push side to side. But in essence these tips amount to minor refinements compared with the mechanical complexity of moving and maintaining equilibrium on a slippery surface. Once the basic technique has been mastered, coaching tips and drills become increasingly important.

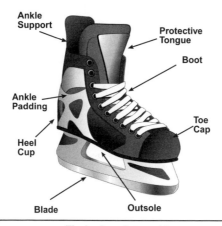

FIGURE 2.1. The hockey skate and its parts

Now let's look at the different parts and functions of the skate and how they help—or hurt—the way the game is played. The skate is divided into two main parts: the boot and the blade (figure 2.1). Let's start with the blade, the first point of contact with the ice.

The Blade

The skate is a unique kind of footwear in that it is designed to optimize motion along one direction and minimize motion in all others. The skate blade enables gliding in both the forward and backward directions, but its sharp edges tend to stop transverse, or sideways, motion. Pushing the skate sideways makes the sharp edges cut and grip onto the ice. The grip is not perfect, nor should it be: if the skate can't drift laterally at all, braking becomes difficult, if not impossible. In hockey, the ability to slow down is just as important as the ability to speed up. This is one of the differences with inline skates (the ones with wheels used to skate on floors) because the wheels don't move sideways. Instead, braking is done with a rubber pad placed at the back of the boot.

Skate Sharpening

The skate blade must be sharp to permit quick acceleration but not so much as to make stopping difficult. The ideal amount of bite on a blade depends on a number of factors, like the softness of the ice, the

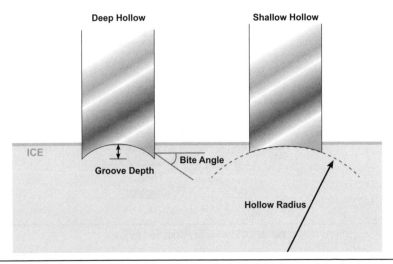

FIGURE 2.2. The underside of a skate blade is hollow; the radius and depth are set during skate sharpening.

roundedness of the blade itself (or the "rocker" seen in figure 1.2), and the weight of the player. The blade's sharpness is controlled by the depth of the groove that appears in the middle of the blade, something called the "hollow," made during skate sharpening. Notice how the blade has two sharp edges and not just one, like knives. Figure 2.2 compares two groove depths on a blade and shows a feature called the "bite angle." Bite angle and groove depth both depend on the grinding stone used on the skate-sharpening machine. Specifically, it depends on the radius of curvature of the outer edge of the circular grinding stone. A grinding stone with a small radius makes a deep groove; a larger radius makes it flatter. Note that it is the roundedness of the edge of the stone that matters, not the size of the wheel itself, as figure 2.3 shows.

One may wonder why skate blades are not sharpened like knives, with a pointy, triangular edge. This shape would certainly provide plenty of grip, but it would also tend to cut too deeply into the ice, slowing down the skater in the process. Having a wide, hollowed bottom enables the ice to support the weight while the blade edges grip the ice at the same time.

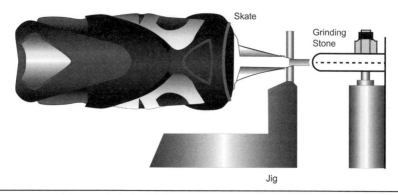

FIGURE 2.3. The rounded edge of the grinding stone creates a matching curvature (hollow) on the blade. In this picture, the grinding stone turns parallel to the blade.

The radius of the grinding stone edge does not all by itself determine how deep the groove is. The width of the blade is important too. The same stone used on a wider blade makes a deeper hollow, and you can convince yourself of this by looking at figure 2.2 again. Regular hockey skates have a blade width of about 3 millimeters, with variations of about 10 percent depending on the manufacturer. Goalie skate blades are wider, at 4 millimeters. The combined effect of the grinding stone radius and the blade width makes it tricky to calculate the groove depth and the bite angle, so I will spare you the details and simply give the results in table 2.1. The table lists commonly used grinding stone radii, expressed in their usual units, which typically range from 5/16 to 1 inch. The most common stone radii (also called "hollows" in the trade, the same term as used above) used for sharpening hockey skates are 3/8 inch (deeper hollow), 1/2 inch (medium), and 5/8 inch (flatter hollow). (I use inches here because it is the unit commonly used to characterize grinding stones.) A deep hollow has pointier edges and tends to dull faster, so sharpening will be needed more frequently.

A flat blade does not dig as much into the ice and this makes sliding more efficient. Some advanced players prefer a flatter hollow, using their skating ability to compensate for the lesser grip at the profit of a better glide, a point I discuss in the following sections. Since heavier players apply more pressure on the ice, they obtain the

Table 2.1. Standard grinding stone edge radii and the resulting hollow depth and bite angle on regular skate blades (3 mm wide) and goalie skate blades (4 mm wide)

Stone edge radius (in)	3-mm blade		4-mm blade	
	Hollow depth (mm)	Bite angle (degrees)	Hollow depth (mm)	Bite angle (degrees)
1	0.04	3	0.08	4
3/4	0.06	4	0.10	6
11/16	0.06	5	0.11	6
5/8	0.07	5	0.12	7
9/16	0.08	6	0.14	8
1/2	0.09	7	0.15	9
7/16	0.10	8	0.17	10
3/8	0.12	9	0.20	12
5/16	0.14	11	0.24	14

same amount of grip with flatter blades as light players get on skates with deep hollows. The technician at the pro shop will sometimes take the weight of the player into account, although the matter can be as much about personal preferences as it is science.

Goalies' needs are a little different. In the T-push technique, a method used by goalies to move quickly around the crease, one foot pushes sideways while the other foot points in the direction of motion, enabling the goalie to glide on it. The grip must be strong for the pushing skate not to slip. But goalies also like to "shuffle," that is, to move from one side of the net to the other without turning their feet, in which case one skate must push and the other skate must drift without difficulty. In the butterfly technique, where both legs extend sideways at once and the goalie falls on his knees, a strong grip would make the trick difficult. Goalie skates are therefore sharpened with a flatter hollow than regular skates.

But again, the most suitable blade sharpening is a matter of taste and style: some use alternatives like non-uniform hollows, with more bite at the front of the skate and less at the back.

You may wonder how the technician at the pro shop uses the same grinding stone on different hollows on different skates. The trick is to "dress" the stone beforehand with a device called a "quill." The tip of the quill is made of the hardest material we know: diamond. Each time a different hollow is needed, the technician reshapes the edge of the wheel by rotating the diamond quill around the edge of the grinding stone, giving it the desired curvature. As amazing as it sounds, diamond cuts through the hard grinding stone like butter, so to speak.

Skate sharpening is a delicate and precise job. The skate is firmly attached to a jig designed to keep the blade parallel to the table and aligned with the grinding wheel (figure 2.3). The jig slides on the table as the skate blade moves along the stone. A dozen or more passes are usually necessary. If the blade is not carefully centered on the stone, the groove will be off center and the blade's two edges will be different. Technicians use a kind of level to check this problem after the sharpening is done. Also, if the skate blade is not kept flat horizontally during sharpening, the groove will move from one side of the blade to the other, with the possible problem of creating a different bite at the front and back of the skate blade.

When there's a problem with sharpening, it's not always the machine's or the technician's fault, however. I once had the recurring problem of losing the back grip of my goalie skates shortly after each new sharpening, a problem I only solved by getting a new pair of skates. As it turned out, the old blades had been warped slightly, so the precisely aligned stone would not have been able to sharpen the blade evenly.

The Glide

The hollow and its effect on the grip is the best-known aspect of skate sharpening. The glide is influenced by factors like ice conditions, humidity, and, perhaps surprisingly, the depth of the hollow: shallow hollows tend to slide easier because they don't have sharp edges digging into the ice, thus causing less internal friction (as discussed in chapter 1).

But more important than the hollow is the roundedness of the blade, something called the "contour" or the "rocker," as seen in the previous chapter (see figure 1.2). The shape of the underside of the blade determines the ideal turning radius (discussed in the following section) as well as the amount of blade touching the ice at the same time, something called the "working surface" or "gliding surface," or simply the "runner." If the skate is upright, between 3 and 5 centimeters of blade typically touch the ice.

So here's the main question: what kind of blade shape produces the easiest glide—a flatter blade or a more rounded one? The answer is somewhat counterintuitive: the flatter blade goes faster. One would think that when more blade rubs against the ice, more friction is produced. But a flatter blade applies less pressure on the ice (less weight per unit of area), so it does not dig as deeply into it, and ice deformation (internal friction) is reduced. Speed skaters use longer, flatter blades for this very reason. Of course, speed skaters would not do well in a game of hockey, where quick turns and maneuverability are just as important as speed.

Weight is another factor to consider here: heavier players need more blade touching the ice to prevent digging too deep. This is why young players glide just as easily as adults do even though their skate blades are very short and rounded.

There are other ways to improve the glide of hockey skates, three of which are worth discussing. The first method is to optimize the geometry of the blade. In one study mentioned in chapter 1, researchers used blades that flared outward at the bottom, thus making the bottom of the blade wider than the top. The widened blades were found to reduce ice friction by up to 30 percent. The better performance appears to originate from the sides of the blade not rubbing against the ice, something that could be significant during a sharp turn. In another study, continual heating of the blade by a lightweight electrical unit (put inside the blade's fixture) to maintain temperature above the freezing point was found to cut friction by 50 percent and to reduce the skater's consumption of oxygen at moderate skating speeds by up to 15 percent. By keeping the blade above the freezing point, a wet layer is formed at the surface of the ice and friction is reduced. Normally, to attain the same effect without heating the blade, one would have to make the ice warmer, say near the freezing point. But as we saw in the previous chapter, although warmer ice is more slippery, it is also softer, which contributes to higher resistance to skating (the optimal ice temperature is around −5 to −10 °C). Skating on warm blades is the equivalent of skating on warm, slippery ice without the drawbacks of softer ice (less dry friction without increased internal friction).

Yet another way to improve gliding is to make blades with materials other than steel, such as titanium and nickel alloys, which have lower friction coefficients. Although some improvements have been made in the area of using alloys, the choice of materials is limited because the material must also offer hardness, rigidity, and resistance to wear comparable to that of steel. It should also be compatible with standard skate-sharpening machines.

Research in the area of improving the glide of hockey skates has produced scientifically proven performance enhancements, but the reception by the hockey community has not always been warm. Innovators wanting to improve the skate glide are not helped by the fact that ice is already quite slippery. In fact, the portion of energy spent by a hockey player to counteract ice friction is typically smaller than that of air friction (assuming a moderate skating speed) and much smaller than the energy spent during the many accelerations, turns, and battles for the puck. As a result, for some hockey players, a reduction in ice friction may go unnoticed. But in racing sports where performances are measured in fractions of a second—speed skating, bobsled, and even downhill skating (ice cross downhill)—a better glide could make a difference.

The Turn

The rocker of the skate blade (the contour radius) owes its name to the rocking chair, as it allows the player to rock his feet back and forth. Of course, rocking is not the player's purpose here but rather having the ability to change direction quickly. Hockey players must continually go around opponents and turn around. Backward skating also relies on easy turning of the skates, as it uses the C-cut technique (the one in which each skate takes turns making C-shaped pushes). As we saw, blade roundedness comes with a price—less glide—but turning is so important in hockey that players must trade the ability to glide for the ability to turn quickly.

The rounded belly of the hockey skate blade is shown in figure 1.2. Note that the blade is not curved the same everywhere, but more so at the front and the back. In other words, the radius of curvature is longer in the middle of the blade than at the heel and toe. But what matters most for turning is the middle portion: when the player skates in a circle, his body leans to the side toward the center of the circle. The circle that most naturally fits the tilted skates is the one with a radius matching that of the blade. Therefore, a sharp turn would benefit from a blade that is more rounded. Of course, this doesn't mean hockey players are constrained to turning in circles of only one radius with a given pair of skates, it only means that some circles are more natural than others.

The rocker radius is mostly a function of the size of the blade. Young players wear smaller skates with smaller rocker radii. Adult skates come with standard-sized blades with rocker radii of about 3 meters (between 9 and 13 feet), while goalie skate rocker radii are closer to 9 meters (30 feet). How could the radius of curvature be much longer than the blade itself? Looking at figure 1.2 again, we see that the radius of curvature is the size of the imaginary circle that best matches the shape of the blade.

Goalies use flatter blades mostly for stability: rocking back and forth would make the goalie unstable. Some players like their blades to be profiled to certain specifications, according to their own preferences and playing styles. To this end, there are machines called "profilers." They operate like a regular sharpening machine, except the amount of steel removed by the grinding stone is not uniform to obtain the desired curvature. Double or multiple contouring of the blade has also been tested with positive results: for example, a flatter contour at the back of the blade can create better push and a more rounded contour at the front of the blade can enhance maneuverability. Some profiling can also be done with a standard skate-sharpening machine provided the operator has the skill to profile both blades the same way.

Blade profiling can also adjust a feature called the "pitch" of a skate. Normally a skate sitting upright on the ice puts the leg in a vertical position. But by making the back of the blade taller than the front of the blade, the skate boot is tilted frontward. This forward pitch puts more of the player's weight on his toes, a more aggressive stance. The pitch can also be set the other way, toward the heels. Manufacturers make certain models of skates that are already set with a forward pitch. Some players find skating with them easier, even backward, and while it facilitates quick acceleration, it's mostly a matter of personal preference, like so many aspects of a player's equipment.

With time, the blade profile tends to change because of wear and multiple sharpenings. Each sharpening removes a thin layer of steel (about a hair thick), which effectively shortens the runner's length, the portion of the blade in contact with the ice. Because each sharpening differs slightly, the left and right skate blades may also acquire different shapes. Not surprisingly, professional hockey players go through dozens of blades every year.

The Boot

The boot of a skate has multiple purposes, one that hockey players should pay as much attention to as their skate's blade.

The first purpose of the skate boot is to firmly hold the foot in place and provide strong ankle support. To make this point clear, let's examine a common injury in sports: the sprained ankle. If an athlete runs on the court, jumps, or makes a quick turn, a sudden torque, or twisting force, is applied to the foot, and the foot can twist inward, putting the outer ankle ligaments under great stress. This torque increases with the distance between the ankle joint and the point of contact with the floor (where the force is applied). The chances of injury depend on the amount of stress applied, the strength of the ankle, and previous history of injury. In hockey, the amount of torque applied to the ankle is roughly twice as much as when wearing basketball shoes on the court. This is because the ankle joint is about twice as elevated from the ice as it would be from the floor with basketball shoes. Without ankle support, hockey players would regularly get ankle sprains. In fact, they would hardly be able to skate at all. Note that the type of ankle support I am talking about here is one that prevents sideways twist, not front-to-back motion, something I will discuss shortly.

Hockey boots have ankle support that rises to about 10 centimeters above the ankle joint. Look back at figure 2.1 earlier in the chapter, which shows the anatomy of a hockey skate. The ankle section includes hard padding and material with a softer texture at the top. The stiffness of the ankle support varies slightly from one model of skate to another. With the heel firmly held in the heel cup, the structure prevents the foot from twisting, especially inward. As a result, sprained ankle injuries in hockey are relatively rare. They more often occur when a player falls and hits the board feet first.

Goalie skates are different in that ankle support rises only to about half that of regular skates. Goalies wear leg pads strapped around the skates that provide additional support to the feet. Thus, ankle injuries among goalies are also rare, but they do happen.

It is important that the foot fits snugly in the boot, with the laces providing a tight hold on the middle of the foot. It is not just a matter of comfort; it also provides a good part of the ankle support. If the boots are too big or the laces too loose, the skater has wobbly feet, as is often seen in children who start skating. On the other hand, tight laces going high up the ankle may cut off blood circulation, the result of which is unpleasantly cold feet and limited mobility. A good habit is to make the laces tighter at the bottom of the boot and gradually release tension near the top.

Some hockey skates are designed with inner moldings that take the exact shape of the feet when the skates are heated, or "baked," to fit the feet of the player exactly. This is an improvement—albeit not a perfect one—over the traditional skates with the one-shape-fits-all kind of boot. Still, most hockey skates are nowhere near as soft and comfortable as the shoes we wear every day, at least not when they are first purchased. It takes time for a new pair of skates to adjust to the shape of the feet, and the feet to adjust to the skates, with appropriately located calluses. After buying a new pair of skates, keep a supply of bandages on hand and expect blisters.

The second purpose of the skate boot is to provide protection against impacts, mostly from pucks and sticks. The front part of the boot is a hard, nearly unbreakable toe cap. The toes need due protection, as they have many of the smallest bones in the foot. The side of the boot also has relatively hard ankle caps, while the rest of the boot and the tongue are made of materials with a softer texture, mostly for comfort. In spite of ankle caps, taking a puck to the ankle is a notoriously painful experience, as any player who's blocked a shot with his feet knows. To reduce the risks of this common injury, some hockey players wear extra padding in the form of custom-made plates that fit around the ankles. In 2013, the Calgary Flames made additional foot protection mandatory following devastating foot injuries to two of their players.

The third function of the boot is to provide ankle motion in the front-to-back direction. This sounds contradictory to what I said about ankle support, but some front-to-back motion is desirable for propulsion. It enables the skater to bend his knees, to lower his center of gravity during skating. This is important, as more power is produced by the legs that way: with bent knees, the legs can contract and fully extend backward (or sideways), thereby delivering the maximum power. A second benefit of frontward ankle rotation is the exploitation of the calf muscles, the ones that raise your heels when you stand on the tips of your toes. These powerful muscles can work through ankle rotation and help propel the skater forward. Speed skaters make full use of their calf muscles with specially designed klapskates, skates with moveable blades on hinges. This design would not work well for hockey players, but improvements can still be made by designing the ankle support carefully. A research group directed by René Turcotte at McGill University found an increase of about 15 percent in power during push off when the ankles are allowed to bend forward and backward by an additional 15 to 20 percent in specially designed hockey boots. This was demonstrated in the 2011 Discovery Channel documentary *Scoring for Science.*

A possible drawback of frontward ankle motion is that when the leg tilts forward, it opens a gap behind the foot and the back of the boot. This exposes the Achilles tendon to the danger of being cut by skate blades, a rare but serious accident. While this can be avoided with specially designed socks or with tendon guards, Achilles tendon injuries still occur (I will discuss them in chapter 6).

We now turn our attention set to another essential part of hockey, the shot and the stick, the subject of the following chapter.

CHAPTER THREE

The Shot

The Stick

Let's face it: hockey players can be very picky about their sticks. Some top scorers consider it their most important piece of equipment, beating out skates and protective gear. After all, it is the only connection they have with the puck, so to have a perfect stick is the key to puck control and accurate passing, shooting, and scoring. NHL superstar Steven Stamkos admits he is superstitious: "If a stick is hot"—meaning that it has scored a few goals—"I won't use it in practice." Another top goal scorer, Phil Kessel, says he can notice when his stick is cut a few millimeters too short or too long.

Mind you, the perfect stick is not always necessary to score a goal. My favorite anecdote about this occurred on December 12, 2010, when Bobby Ryan, then playing for the Anaheim Ducks, scored a goal against the Minnesota Wild with a stick that had the wrong blade curve. Ryan shoots right, and the scoring stick was for left-handed players. Actually, it wasn't even his stick. What happened was a rare event: Minnesota player Mikko Koivu had dropped his stick during a corner-ice scramble, and, instead of picking it up off the ice, he skated over to Bobby Ryan, and with the speed of a magician, stole Ryan's stick from his hands (of course, he had no time to check whether it had the right blade curve). Surprised, Ryan raised his hands to claim a penalty, but the referees had seen nothing. So he picked up Mikko's stick lying on the ice. As play continued, he received a pass and scored on a nice one-timer (an immediate shot upon reception of the pass). Of course, the Wild's coach protested that the goal was scored with an illegal stick, but the officials ruled it a good goal. (You can find a video of this online by searching for "Bobby Ryan Koivu stick.")

The hockey stick, which began its life as a plain old piece of wood shaped like an L, has gone through several makeovers throughout history. And the pace of innovation seems to be accelerating, thanks to technology and engineering. Sticks now come in a wide variety of lengths, blade curves, types of material, and shaft stiffness and sizes. And, as usual, more choices mean harder decisions. In a single chapter, I would not be able to cover all or even most of the recent developments in hockey sticks, so I focus instead on key features, including shaft flexibility and blade shape, which are known to affect shooting speed and accuracy.

Stick Length

Stick length is often the first criterion players use when selecting a stick. The general consensus is that when holding the stick vertically, with the tip of the blade touching the ice, the end of the shaft should come up to around the player's neck. When a player is not wearing skates, the stick will then come up to around his nose. Of course, a longer stick can be selected and be cut to the desired length.

There is not a whole lot of science behind this rule of thumb, and there is certainly room to suit a player's style. An offensive player may prefer a shorter stick for better puck handling; it's harder to maneuver a puck up close with a long stick. Coaches recommend shorter sticks to young players for the same reason and also because shorter sticks force the good habit of assuming a lower, more stable posture on skates. Defensemen, in contrast, may want a longer reach to intercept passes. There is a limit, however: the official NHL rulebook does not allow sticks longer than 63 inches (160 centimeters) from the heel to the end of the shaft but will grant extensions of up to 65 inches (165 centimeters) to some players who are 6'6", or taller.

The Flex

The flex is one of the most discussed properties of hockey sticks. Before we look at how flexibility affects shooting in hockey, we need to examine some technical aspects of the flex.

When pushing the middle of the shaft while holding the end with the other hand (figure 3.1), the shaft offers resistance that increases in proportion to how much the stick is bent. In other words, if the

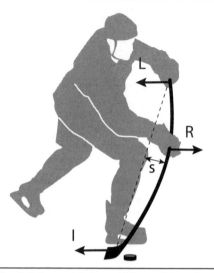

FIGURE 3.1. A right-handed shooter bends a stick by applying a force near the middle of the shaft with his right hand (R), which transfers half this force to the left hand (L) and half to the ice (I) and the puck. The shaft bends and sinks by an amount *S* near the middle.

applied force *R* on the stick is doubled, the sinking of the middle of the shaft (the quantity *S* in figure 3.1) also doubles. The flex rating is a number that normally ranges from 30 for youth player sticks to about 110 for the stiffest adult hockey sticks. The number is meant as an indication of the amount of force *R* (in pounds) that must be applied to sink the middle of the shaft by 1 inch. In scientific units, an adult stick with a medium flex rating of 75 offers 130 N of force for each centimeter of sinking at the middle, so it takes 260 N to sink it by 2 centimeters, and so on. On some of the stiffer commercially available sticks, this number is as high as 200 N per centimeter of sinking. Although professionals will often use stiff sticks, figure 3.2A shows how bent a stick becomes at the peak of a slap shot.

It is interesting to note that the amount of force transmitted to the blade (and the puck) is only half of the force applied at the middle of the stick. This is a result of the laws of physics applied to levers.

An important property of the flex is that it depends on where the force is applied on the stick. By putting your hands at different positions on the shaft, you can find out that it is harder to bend a short

FIGURE 3.2A. (*left*) Buffalo Sabres' Drew Stafford flexes his composite stick during a slap shot. At the peak of impact, the stick can bend by as much as 20 degrees. (AP Photo/Gary Wiepert) **FIGURE 3.2B.** (*right*) On this quick wrist shot by Nicklas Grossmann of the Philadelphia Flyers, the puck was shot from 25 feet away, leaving too little time for New York Rangers' goalie Henrik Lundquist to react, beating him in the top left corner of the net. Notice how the shooter's back leg is used for equilibrium while he puts some of his weight on the stick to bend it before the puck is released. (Kostas Lymperopoulos/CSM, Cal Sport Media via AP Images)

section of the stick than it is to bend a long one. This is why the flex rating of a stick increases when it is cut shorter. Some manufacturers draw lines on the shaft and write numbers corresponding to the modified flex in case a cut is made there.

Another consequence of this is that the amount of stick bending depends on where you hold the stick during a shot. Putting one hand lower, closer to the blade, increases the force applied on the blade, in the same way a lever applies more force to a load when the fulcrum is put closer to it. During a slap shot, most players put their pushing hand near the middle of the shaft, a little bit closer to the blade, but some put their pushing hand lower in an attempt to deliver more force to the puck.

When hockey fans watch their favorite NHL star, like Shea Weber or Zdeno Chara, fire pucks with sticks that are very stiff, they could easily conclude that the stiffer the shaft, the faster the shot. Some of the bigger and stronger NHL players, such as Hal Gill, use sticks with flex ratings of 120 and higher. On the other hand, snipers like Phil Kessel and Alexander Ovechkin use softer sticks with flex in the 80s. Tests done in a controlled laboratory environment suggest that there is no general rule that applies to all players.[1] The average puck speed showed little or no change when sticks of different shaft stiffness were used by a group of players. The performance of a single player testing different sticks showed less variation between sticks than the shot-to-shot variation in puck speed with the same stick.

The relative insensitivity of slap shot speed to the shaft stiffness agrees with a simple physical model I discussed in my earlier book, *The Physics of Hockey* (Johns Hopkins University Press, 2002). Let's recap it here for those who haven't read it. The slap shot can be understood as a rotating object (the player and his stick), followed by the bending of the stick when it hits the ice and the release of the stored energy to the puck. Such elastic interaction conserves kinetic energy. For example, studies have shown that during the loading-unloading cycle of a hockey stick, the shaft returns 90 percent of the energy stored in the flex.[2] This energy, which comes from the rotation of the player (not from the stick itself), is the determinant factor in how much energy the puck receives. In other words, the amount of stick bending is not the indicator of puck speed; what matters instead is the mass and rotating velocity of the player during the slap shot, or the original kinetic energy.

When the slap shot is properly executed, the blade of the stick hits the ice just before the puck, and the resulting flexing of the stick stores energy into it. This energy is coming from the motion of the player, who slows down in the process. Later, the blade comes in contact with the puck, transferring the energy to it, pushing it forward. This is the "whipping" effect. The more flexible the stick is, the more bending there will be, and the whipping (acceleration) of the puck will be weaker but will be taking place over a longer distance.

1. J. T. Worobets, J. C. Fairbairn, and D. J. Stefanyshyn, "The Influence of Shaft Stiffness on Potential Energy and Puck Speed during Wrist and Slap Shots in Ice Hockey," *Sport Engineering* 9 (2006): 191–200; D. J. Pearsall, D. L. Montgomery, N. Rothsching, and R. A. Turcotte, "The Influence of Stick Stiffness on the Performance of Ice Hockey Slap Shots," *Sport Engineering* 2 (1999): 3–11.

2. Worobets, Fairbairn, and Stefanyshyn, "The Influence of Shaft Stiffness on Potential Energy and Puck Speed."

While the puck speed is about the same for different stick stiffnesses, a stiffer stick produces a greater force at the peak of impact. For some players, a stiffer shaft may produce an unpleasant stinging sensation in the hands. (Readers may wonder why greater force doesn't automatically translate into higher puck speeds. This is because puck speed is determined by how much work is done on it, and work equals force multiplied by displacement. A stiffer stick produces more force over smaller displacement, and the end result is the same as when using a softer stick.)

The wrist shot is different. A technique that is more accurate, quicker to release than the slap shot, and about 30 percent slower, the wrist shot is more like pushing a puck and less like hitting it. The pushing comes from the player sweeping the puck, and the whipping comes from the bending of the stick as the player purposely pushes it against the ice during the sweeping motion. Players who truly master the wrist shot exploit the whipping effect to its maximum (as in figure 3.2B). For both the slap shot and the wrist shot, about one-third of the energy given to the puck comes from the flexing of the stick and the resulting whipping effect. The difference, however, is that the wrist shot benefits more from the flexing. Studies have shown that wrist shots tend to go faster when using softer shafts (lower flex ratings and more bending of the shaft). Players seem to be able to store more energy in a softer stick. But the difference is not enormous: compared with sticks rated "very stiff," sticks rated "medium" produce wrist shots that are 5 to 8 percent faster.

In general, a player should choose a stick with a flex as soft as possible while keeping in mind that a lot of shaft bending makes it harder to produce consistent, accurate shots. The other obvious problem is that a bending weak shaft could simply break, destroying the (often expensive) stick. So, by and large, the choice of flex is above all a function of the player's size and strength, and this is the reason that flex ratings start at the bottom with children's sticks and end at the top with professional adult sticks. Professional players can afford to use sticks with stronger flex, and, if their shots go faster, it is not so much because of the stick but due to the fact that they are stronger, are heavier, and use a well-tuned technique.

On the subject of shooting technique, research has revealed a number of subtle differences in the way elite players and recreational players[3] use their sticks and body weight during wrist shots and slap shots. Even when the two groups were similar in strength and size, the elite players had faster and more accurate shots. For example, during a wrist shot, they had a tendency to lock their shoulders and make the adjustments with their wrists, which appears to be more consistent, accurate, and effective. Accurate shooters also move their center of mass toward the front foot, extending their back leg to keep their balance (as in figure 3.2B). This ensures body stability throughout the motion, a necessary condition for accuracy and shot repeatability.

What Are Sticks Made Of?

For a long time, hockey sticks were made by gluing pieces of wood and other materials together. Now, thanks to composite technology, some sticks are produced as one solid piece from one end to the other. Composite sticks are made from materials like fiberglass, carbon fibers, and graphite mixed into a hardening resin. The fabrication is generally a molding process. Some composite sticks are made with detachable blades, like the old aluminum sticks. An advantage of these sticks is that it gives a choice of blade while keeping the same shaft we are accustomed to. Composites offer new possibilities for molding blade shapes and curvature as well as tweaking stiffness along the shaft. They thus offer a wider range of flex than what was traditionally available with wood and aluminum. Composites are also lighter and more durable than wood is, in the sense that they tend to retain their mechanical properties longer. They have a different feel, too, which is something even star players like Sidney Crosby had to get accustomed to when they switched from wood to composite sticks.

As far as performance goes, a bent composite stick tends to release its energy more efficiently. There is less damping and less absorption of energy, creating a different feel to the hands. For this reason, when

3. K. V. Lomond, R. A. Turotte, and D. J. Pearsall, "Three-dimensional Analysis of Blade Contact in an Ice Hockey Slap Shot, in Relation to Player Skill," *Sports Engineering* 10 (2007): 87–100; T.-C. Wu, D. Pearsall, A. Hodges, R. Turcotte, R. Lefebvre, D. Montgomery, and H. Bateni, "The Performance of the Ice Hockey Slap and Wrist Shots: The Effect of Stick Construction and Player Skill," *Sports Engineering* 6 (2003): 31–40.

the stick is flexed, a greater percentage of its stored energy is transferred to the puck. The differences, on the order of 10 percent, are not enormous, however.[4]

Because some composite sticks carry a higher price tag, it should be stressed that performance enhancement is not guaranteed to all players. When a player is still learning to master the different techniques of shooting, expensive composite sticks are probably not needed. (That's why I won't buy a set of costly golf clubs—I know it won't improve my game.) Many amateur players shoot the same whether they use wooden or composite sticks.

The Goaltender's Stick: Wood versus Composite

While the vast majority of players have switched to composite sticks, some goalies have held on to their wooden sticks. Some of them use composite sticks because they weigh less, but others favor wood. Part of the reason is that wood tends to dampen vibrations more. When a puck traveling fast hits the goalie's stick directly, the sting of vibration of the wood stick is less than it is with a composite stick. This is also the reason that a wooden stick is (slightly) less efficient in a slap shot and why it feels different when handling the puck: the knocking vibrations are absorbed and aren't transmitted as much along the shaft to the hands.

The difference in the way that vibrations move along sticks of various materials was verified in a scientific study, using precise vibration measurements.[5] This study showed the regions of the stick that remained practically stationary and the other parts that vibrated intensely. It is a wave similar to the one found in the nodes and antinodes of a guitar string that is vibrating. If the goalie happens to hold the shaft where it does not vibrate a lot, his hand will not feel much of anything when a puck hits the stick.

Researchers found vibrating frequencies of composite sticks to be higher than that of wooden sticks, owing to the fact that composites are stiffer and lighter. These higher frequencies fall in the 200 to

4. D. J. Stefanyshyn and J. T. Worobets, "Energy Return and Puck Speed of Hockey Sticks," Abstracts of the 5th World Congress of Biomechanics, *Journal of Biomechanics* 39 (2006).

5. L. J. Hunt and D. A. Russell, "Vibrational Assessment of Ice Hockey Goalie Sticks," *Journal of the Acoustical Society of America* 130, no. 2429 (2011).

400 Hz range where human hands are most sensitive to vibration stimuli. By putting a mass of tape at the end of the shaft, as goalies do for safety reasons and to not lose hold of their sticks, vibrations were dampened and the frequencies lowered, thereby reducing the sting. Also, the nodes (the locations with no vibrations) were shifted along the shaft, which would be useful to reduce the sting.

The Shaft Size

The size of the shaft is an important consideration when making the transition from a youth to an intermediate to an adult (senior) stick. A larger shaft makes it easier to apply more torque to the stick and to the blade. During a slap shot, the puck pushes against the blade and twists the shaft, and with a good grip, the hands prevent that from happening. The same is true when receiving a hard pass; the player needs to avoid having the stick rotate as the blade receives the puck, acting as a lever. It is best for younger players to use smaller shafts, however, as it allows for easier stick handling.

Some shafts have rubber-like wrappings to provide a better grip. Some players tape the shaft in a criss-cross pattern and others add sticky fabric to the inside of their gloves to get a better grip.

The Curve of the Blade

I could have started this chapter with the stick blade because it is often the first feature hockey players look at in a stick: first to check the stick's handedness (right-bending blades for left-handed players and left-bending blades for right-handed players) and then to examine the type of curvature. The blade is important, as it is the point of contact between the player and the puck, so shooting and passing accuracy depends on it.

The stick blade is a surface that is both curved and twisted, so it cannot be described in just a few words or numbers. Nonetheless, there are some key features that need to be considered, starting with the depth of the curve. The blade is slightly U-shaped, as figure 3.3 shows.

It is a common misconception that hockey players historically switched from flat blades to curved blades because they produce faster shots. The curve is mostly about puck control and consistency of shots. The blade's curve brings the following three improvements:

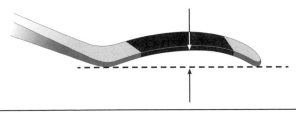

FIGURE 3.3. The curve depth of a hockey stick blade is the distance between the arrows.

- *Consistency*: The curve effectively forms a pocket at the bottom to which the puck will tend to go. If the puck leaves the blade near the same place each time, the shot is more consistent and accurate.
- *Control*: It is easier to scoop the puck and move it around an opponent with a curved blade. Other tricks are also made easier, like grabbing the puck at the tip of the blade and shooting it upward all in one move, the so-called toe drag.
- *Puck spin*: It can hardly be seen by eye (slow-motion videos are best to see the effect), but a curve permits more puck spin (shown in figure 3.4 for a left-handed shooter). A spinning puck, like a spinning football, is more stable. Without it, the puck flips in the air and loses speed as its movements become erratic. The spin is partly created with a flick of the wrists during shooting. In a "saucer pass," where the puck is shot in the air between players, the puck spin is especially important because the puck must land flat on the ice for easy receiving. Although it is technically possible to spin the puck with a straight blade, it can be done better and more consistently with a curved blade.

There is an NHL rule limiting the curvature depth to 1.9 centimeters, and this is probably to avoid excessive puck control. To verify the legality of the blade, some use the "dime test" (not quite 1.9 centimeters but close) whereby the coin should just barely slip vertically underneath the blade when the latter is lying against the floor. NHL referees have measuring devices to control illegal sticks. Note that the current limit was an increase from 1.3 centimeters as of 2006.

Some years ago, I half-jokingly made the suggestion that without a limit on blade curvature, hockey stick blades would evolve into rounded devices capable of firmly holding a puck. The same idea was

FIGURE 3.4. As the puck rolls along the blade during a shot, it acquires the spin needed for stability and accuracy.

later used in a humorous Molson Canadian beer commercial, where an actor uses the same kind of stick to pick up a bottle of beer from a fridge without leaving his couch.

A second feature of the curve is the location on the blade where it begins. A blade can be curved like a circle, uniform and starting at the heel. But, as shown in figure 3.5, some blades have curves starting near the heel and others near the middle or the toe of the blade (the blade on the extreme right). With a toe curve, it is perhaps a little easier to pick a puck away from someone else, but, by and large, the different types of curves are a matter of preference. Once players are accustomed to shooting a certain way with a certain blade type, they often like to keep the same blade. For this reason, stick companies like CCM, Bauer, and Easton provide custom-made stick blades for NHL players, each player having his own preferred blade pattern.

Another aspect of the blade shape is the "loft" or "face." The loft is the tilt angle of the blade: you can see it when holding the stick and looking from above. A blade that tips backward is said to be more "open faced." If you know a bit about golf, you can imagine this as being similar to the difference between a nine iron and a three iron. As in golf, the more tilt a hockey stick has, the easier it is to shoot the puck into the air.

The toe shape is another aspect of the blade. The toe is the very tip of the blade, and it comes in round and square shapes (as seen in the left blades of figure 3.5). A square toe offers perhaps more blocking area and the round one gives more puck control at the tip, like during a toe drag.

Finally, the "lie" (figure 3.6) is the angle the blade makes relative to the shaft. It is given by a rating ranging between 4 and 8 that is

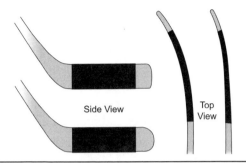

FIGURE 3.5. Examples of stick blade curvatures and profiles. *Top view*: the blade on the right has a "toe curve," and the one on the left has a "heel curve."

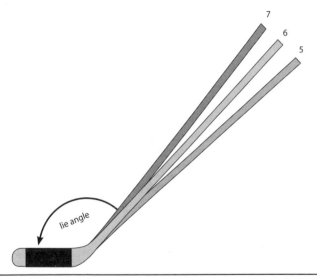

FIGURE 3.6. The lie of a stick is the angle between the shaft and the blade. The angle typically ranges from 140 to 150 degrees. Sticks with higher lie numbers are more upright.

printed in front of the shaft. The higher the number, the more up-right the shaft is. For the proper lie, the bottom edge of the blade should be flat against the ice when the player is holding the stick as he would normally.

What about the negative aspects of a curve? There are not that many, to be honest. But if a curve helps the consistency of a for-ward (forehand) shot, it hurts the backhander. It is harder to have a consistent backhander when the puck has a tendency to roll away from the center. Another problem with some blades is that their finish is rather slippery and doesn't provide grip to spin the puck properly. This is why players wrap their blades with cloth tape, not just to protect them, but to make for better puck control and spinning. Alternatively, the same effect is achieved with spray-bottle products that make the blade more sticky and rough. The crosshatched side area of the puck also helps to create the needed friction to spin it. The most common color of tape is black, which adds the slight advantage of hiding the puck from the goaltender, as opposed to white tape, which enhances the contrast between the puck and the blade. According to the NHL's official rulebook, "Adhesive tape of any color may be wrapped around the stick at any place for the purpose of reinforcement or to improve control of the puck."

Handedness in Hockey

There are more left-handed than right-handed shooters at all levels of hockey. In hockey, "left-handed" means the shooter holds the stick with the blade to his left. And manufacturers know this, as they sell more left-handed sticks. Table 3.1 divides skaters who played at least one game in the 2010–11 NHL regular season by handedness. In all, two-thirds of players shoot left. Goalies are even more lopsided in favor of the left: 90 percent of NHL goalies catch the puck with their left hand and therefore shoot left.

At first, these numbers are surprising, for two reasons: first, in the population as a whole, only 10 percent of people are considered left-handed; second, there is as much demand for right-handed as left-handed players, in both defense and offense, so a 50-50 distribution would be ideal and make sense.

To explain this puzzle, let's start with the simple fact that the dominant hand, the one that is more "skilled," is the one we like to use for difficult tasks. For example, it is easier to throw a ball fast and accurately with the dominant hand, so baseball players typically use their non-dominant hand to catch balls. In hockey, the dominant hand is usually best placed at the butt end of the stick. This might sound strange at first, as one could think of naturally picking a stick at the middle with the dominant hand. But in a game, when a hockey player holds his stick with one hand, for example, as he extends his

Table 3.1. Handedness in hockey according to position played

	Left wing	Center	Right wing	Defenseman	Percent of total
Shoots left	161	169	37	190	62
Shoots right	15	79	135	107	38

Source: Data from the 2010–11 NHL season.

reach in a poke check, this is best done with the dominant hand holding the stick at the end. Also, during a wrist shot, much of the stick-twisting action is done by the hand at the end of the shaft, the other one used mostly for pushing. As a result, since the majority of people are right-handed, they end up shooting left in hockey.

In terms of playing position, left-preferring shooters tend to play on the left wing, regardless of whether they play offense or defense, while centers can shoot from either side. There are exceptions, like Washington Capitals' Alexander Ovechkin, who shoots right, but has been playing on both wings.

It is interesting to note that there seems to be a far greater proportion of left-handed golfers in Canada than in the rest of the world, both at the amateur level and among professionals (like PGA player Mike Weir). Some have attributed this phenomenon to the popularity of hockey in Canada, a country where most children learn to handle a hockey stick before they learn golf, so they naturally swing left when picking a golf club. This is not surprising in a country where summer sports are often considered to be activities just to stay in shape for the following hockey season.

In terms of effectiveness, both left- and right-handed forwards appear to perform the same in the NHL, as table 3.2 suggests, with both having practically the same points per game. There is a slight difference for defensemen, however. The difference is statistically significant and repeats itself year after year: right-shooting defensemen score 13 percent more points per game than left-shooters. Since both groups get roughly the same ice time, the reasons for this difference are unclear. Perhaps right-shooting defensemen get more time on power plays, as having both right- and left-shooters on the ice at the same time may be advantageous. In any case, right-handed

Table 3.2. Points per game by left- and
right-shooting players in the NHL

	Forward	Defenseman
Shoots left	0.453	0.281
Shoots right	0.457	0.320

Source: Data from the 2010–11 NHL season.

defensemen are a hot commodity in the NHL. In principle, every
NHL team would like to have three right-handed and three left-
handed defensemen on its roster, but only 36 percent of them shoot
right.

The main benefit in having right-handed defensemen is that they
can play on the right wing and go hard to the boards in their own
zone to make a clearing attempt with their forehands. When a left-
handed defenseman has to play the right side, he has to clear the puck
along the board with his backhand or turn around to clear it, both
of which are tricky and slower. In the offensive zone, when the puck
is passed along the boards, it is harder when the defenseman has
to rotate to gain possession of the puck or receive the puck with a
backhand.

The Bounce Effect

I was once asked by *The New York Times* to explain a seemingly puzzling phenomenon in hockey. Sheldon Souray, then the ruling king of
the NHL's Hardest Shot competition, an event held at the NHL
All-Star Skills Competition, had said to the *Times* that he thought
that his slap shot went a little faster when he executed a one-timer.[6]
In a one-timer, the player slaps the puck upon reception, without
stopping and controlling the puck first. It takes a lot of skill and
perfect timing, but the trick leaves little time for the goalie to
prepare.

There was little reason to question Souray's instinct as he twice
scored more than 20 goals in a season, a rare feat for modern defensemen. At the 2004 All-Star meet, his slap shot had been clocked at

6. J. Z. Klein, "Rubber Bullet," *The New York Times*, March 4, 2007.

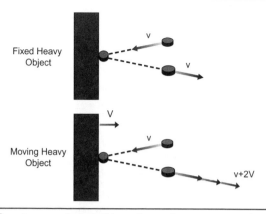

FIGURE 3.7. The return speed of a ball bouncing against a more massive object is increased when the massive object is moving toward the ball.

102.2 mph, tied with Adrian Aucoin's as the winner of the Hardest Shot contest, so he knows a thing or two about how to send a puck flying. Still, our gut feeling tells us that hitting an oncoming puck, like running uphill, would be counterproductive and more difficult. Let's see if physics agrees.

As we know from experience, a rubber ball thrown against a heavy object (like a wall) will bounce and come back at almost the same speed. Figure 3.7 illustrates the idea. But if the heavy object were moving toward the person throwing the ball, would the ball come back at the same speed? According to the laws of physics (the principle of conservation of momentum, to be specific), the ball would bounce back with the speed it had initially plus twice the heavy object's speed.[7] For this to happen, the collision must be elastic (recall this means no loss of energy and no permanent deformation).

Let's consider three examples of this effect in sports. During a golf drive, a ball resting on a tee is hit by a driver. In this situation, the ball has an initial speed of zero and the heavy object, the head of the driver, has some speed. According to the aforementioned principle, the golf ball should acquire twice the driver's speed, and indeed this is what it does. (In reality it may be slightly less than twice the driver

7. Mathematically, the ball bounces back with a velocity $v = v_1 + 2v_2$, where v_1 is the ball's incoming velocity and v_2 the heavy object's velocity.

speed because the driver is not hugely more massive than the golf ball.) Another example is tennis: during a return, the ball is impacted by an oncoming racket that sends it back at a greater speed than it initially had. A final example is found in baseball. The pitcher throws a ball, which later collides with the much heavier bat, sending the ball at a greater speed than both the bat's and the initial throw's. A home run is, in a way, the combined effort of the pitcher and the batter (it would be harder for the batter to score a home run by hitting a stationary ball). For all three examples given, slow-motion videos showing the bounce effect can be found on the Internet.

Executing a slap shot on an oncoming puck is in many ways similar to hitting an oncoming tennis ball with a racket. The stick and the player are many times heavier than the puck, so it would bounce with the same speed as it was received in addition to acquiring the speed it would normally receive by the slap shot. Of course, the oncoming puck speed is relatively modest—it would add perhaps a few miles per hour, as passes are much slower than slap shots—but physics would agree with Sheldon Souray's observations.

The Puck

The hockey puck is a cylinder of solid rubber with the standard measurements of 1 inch in thickness and 3 inches in diameter. The puck was not always shaped like a disk (it looked more like a ball originally) but it evolved to be so probably out of the necessity of keeping it close to the ice. There is a sport called "bandy" that is similar to hockey but is played with a round rubber ball, and, perhaps for that reason, the nets are almost as big as soccer nets.

The puck has a mass of about 6 ounces (170 grams), making it denser than water (which only really matters if you try to retrieve your puck on a pond after it has thawed). It has a crosshatch texture all around the edge to increase the friction with the stick blade for better puck handling and spinning. Hockey pucks are made of vulcanized rubber and are colored black for better contrast against the ice. Vulcanization is a chemical process in which the rubber is hardened and rendered more durable with a treatment using sulfur and high temperature. A puck is very stiff, and unlike golf balls and baseballs, it does not compress very much when hit at full impact, at least not with a hockey stick.

Tests made with a hydraulic press showed that to compress a puck by 10 percent (along the 3 inches of its horizontal axis), a force equivalent to the weight of a 180 kilogram mass would be needed. This is 5 times the force applied to the puck during a hard slap shot, in which it compresses by about 1 millimeter. If the same puck travels at 80 mph and hits a hard wall, it squeezes by about 2 centimeters. It is a rare event, but pucks have been broken in half during a game. When United States Hockey League (USHL) defenseman Andrew Prochno did a routine one-timer slap shot from the blue line, the puck hit the side post of the net and split in half. None of the pieces entered the net, but the rule states that the entire puck must cross the goal line to count as a goal. During a practice with the Eerie Otters in 2014, Connor McDavid made a puck explode into several pieces when it hit the crossbar. Flaws in the rubber are probably to blame for such breakage.

In the NHL and other hockey leagues, pucks are kept frozen prior to a game to make them less bouncy and able to slide better on ice. Warm pucks stick more to the ice and bounce more off the ice and the boards, and so are harder to control.

To improve their shooting power and stick handling during practice, hockey players sometimes use heavier pucks that weigh typically 10 ounces (280 grams). It's good resistance training since the effort increases proportionally to the mass to send the puck flying at the same speed.

There are also a variety of foam pucks used in friendly games, where players are not fully equipped, or for playing indoor hockey and for protecting walls and windows. Foam pucks are safer, but they don't slide very well on ice and lose their speed more quickly when shot in the air. During the 2014 Ottawa Senators' Skills Competition, hulking defenseman Jared Cowan fired a puck at 110.5 mph, better than the fastest shot ever recorded by an NHL player. For a few moments, he thought he was the new record holder, but because someone had replaced the legal puck with a foam one, it did not count. It was all in good fun. The prank was meant to create a puck that fluttered, as foam pucks often do, so everyone was taken aback when the radar revealed the blistering speed.

During a slap shot, the lighter the puck, the faster it goes, but only up to the limit set by the whipping speed of the stick blade. Note that at 30 grams (1 ounce) of mass, foam pucks are many times lighter than regular pucks. Because of air drag, the foam puck slows down much more quickly than a regular puck (I will examine air drag in the following section). When shot from the blue line, a foam puck loses one-third of its speed by the time it reaches the net.

Aerodynamics of a Hockey Puck

Watching hockey players shoot pucks at great speeds, we don't usually think of the puck as something that would be greatly affected by air. It's not like a Frisbee, which is carried long distances by air lift. Yet aerodynamics does influence a puck's trajectory, as we will see. The aerodynamics of a puck has been measured in some detail in the context of a study on spectator safety. It was done to determine how tall the windows placed around the rink boards should be to protect spectators from pucks gone astray.[8]

In all, there are four aerodynamic forces acting on a puck shot in the air:

1. A drag force pushes against the puck in the direction opposite to its movement, slowing it down.
2. A lift force acts vertically, helping to keep the puck in the air.
3. A "spin-down" force slows down a puck's spin.
4. A "pitch moment" force tends to flip the puck onto itself.

Figure 3.8 shows these aerodynamic forces. All the forces are manifestations of the same effect, namely air pushing and rubbing against the puck, and all four forces become more important as puck speed increases. In fact, aerodynamic forces increase with the speed squared; doubling the speed quadruples its effect.

A puck experiences a drag force equivalent to half its weight when it is traveling at 90 mph. Down to 70 mph, the drag force drops to one-third of the puck's weight. To give an idea of how this affects the puck's speed, for initial speeds between 60 mph (a good wrist shot) and 100 mph (a hard slap shot), the puck loses 5 percent of its original speed because of air when going from the blue line to the

---

8. H. Bohm, C. Schwiewagner, and V. Senner, "Simulation of Puck Flight to Determine Spectator Safety for Various Ice Hockey Board Heights," *Sports Engineering* 10 (2007): 75–86.

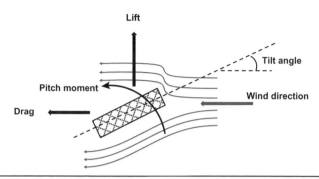

FIGURE 3.8. Drag, lift, and pitch moment (torque) are three of the four aerodynamic forces acting on a moving puck. The tilt angle is measured relative to horizontal. Unequal amounts of wind pressure around a tilted puck result in a torque that tends to flip the puck around itself.

net. The same puck traveling from one end of the rink to the other would slow down by 15 percent.

When a puck travels along the ice, air drag is not the same as when the puck travels in the air. In a flight, air flows around, on top, and underneath the puck, whereas against a surface, air can only go on top and around the puck. When air has fewer options to move around the puck, air friction is increased, so a puck moving along the ice experiences 16 percent more air drag than when it travels through the air. Interestingly, a puck moving flat along the ice also experiences an upward force caused by a higher wind speed on top of the puck (a horizontal puck in the air does not experience lift because the air flow on top and below is the same). This creates a suction force that is similar to the lift force on an airplane wing, although it is much more efficiently done on a wing. Still, at 100 mph, this lift force represents two-thirds of the puck's weight, and this contributes to reducing the ice friction force by the same amount.

If the puck is tilted, air drag is increased in proportion to the puck's increased cross sectional area. With a 20 degree tilt, the puck velocity drops by 30 percent from one end of the rink to the other, and by 12 percent from the blue line to the net (for shooting speeds between 60 and 100 mph).

There is also a phenomenon called the "drag crisis." Beyond a certain speed, air turbulence around an object causes air drag to increase not as the velocity squared but something more linear. With a baseball, this effect begins at speeds of 60 mph and levels off at 100 mph, speeds well within the reach of pitchers and batters.[9] Because the hockey puck is about the same size as a baseball, it is expected that the drag crisis would occur at around the same speeds, but this has not yet been precisely measured.

On slow-motion replays of slap shots, a puck that is tilted upward sometimes appears to glide over the ice. This is the result of an upward force canceling out gravity. Even for small tilt angles, say between 3 and 20 degrees, and puck speeds typically encountered in hockey, the aerodynamic lift is sizeable.

There are two questions of interest here:

1. At what speed does a puck become airborne?
2. Even if the puck is not completely airborne, does aerodynamic lift affect the puck trajectory?

From measurements made in wind tunnels, it was found that aerodynamic lift equals the puck's weight at speeds between 32 and 38 meters per second (between 70 and 85 mph) for angles of attacks (tilt) between 5 and 15 degrees. The puck would then be seen as "floating" in the air, being airborne. But at these speeds, it reaches its destination quickly, so we have no time to really appreciate its graceful glide in the air.

When I was a child, I had a coach who, for fun during practices, would fire a puck from one end of the rink to the net at the opposite end, the puck never touching the ice in between and rising to about 2 meters or so above it. As a youngster, I watched this in awe. As it turns out, this feat would be impossible without some aerodynamic help. In a vacuum, such low trajectory would require a launch speed of 50 meters per second (110 mph) or more. But in the real world, with air, the required shooting speed is about 30 meters per second (less than 70 mph), a speed within reach of amateur players. Figure 3.9 shows drastic differences in puck trajectory when there are aerodynamic forces and when there are none. Without air, the puck falls back to the ice quickly, then slides to the net. With air and a small tilt angle, the puck is carried up much farther, like a Frisbee.

9. R. K. Adair, *The Physics of Baseball*, 3rd ed. (New York: HarperPerennial, 2002).

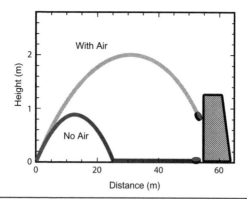

FIGURE 3.9. Puck trajectories, calculated with air and in a vacuum without air, for a puck shot from one end of the ice to the opposite net. The puck travels at 30 m/s (68 mph) and is launched at an angle of 8 degrees from horizontal. The attack (tilt) angle is also 8 degrees and is assumed to remain the same throughout. For clarity, the aspect ratio in the figure is not 1.

Shots on goal are usually taken from distances typically no farther than the blue line, some 20 meters away. Does air affect the puck trajectory in such a case? Let's consider a 38 meters per second (85 mph) slap shot taken from the blue line and aimed at the top corner of the net (1.2 meters high). Without air, the shot aimed at the vertical bar would return to the ice before it reached the net. But with aerodynamic lift, it hits the net at the center. So air does help the shooter on close shots too.

Let's now turn to puck spin. The stick blade usually imparts a spin on the puck when it is shot. I discussed the importance of this effect on stability in the previous sections. The rate of spinning is typically between 10 and 30 rotations per second (as measured in a laboratory). At that rate, by the time the puck reaches the net from the blue line, it will have spun 20 times or so.

Air friction slows down that spin, but measurements suggest that the effect is weak: it would take about a minute for a puck to lose half its spinning speed because of air. Pucks never travel in the air for more than a couple of seconds, so we can assume the spinning to be constant during that time.

Finally, let's examine something called the "pitch moment." Pitch moment can be seen in figure 3.8. In that figure, air flow around a

Table 3.3. Maximum distances traveled by a puck with ice and air friction

Initial velocity (mph)	Ice friction only		Ice and air friction	
	Distance traveled (km)	Travel time (min:sec)	Distance traveled (km)	Travel time (min:sec)
50	0.5	:45	0.25	:29
60	0.7	:50	0.31	:31
70	1.0	1:00	0.36	:33
80	1.3	1:10	0.41	:34
90	1.6	1:20	0.45	:35
100	2	1:30	0.49	:36

tilted puck is asymmetric: the path going over the puck is not the same as the one going under it. This causes unequal amounts of wind pressure on the top and at the bottom of the puck, the overall result being a torque, a pitch moment, which makes the puck flip backward onto itself. The only time there is no pitch moment is when the puck is oriented at zero or 90 degrees, in which case the air paths are symmetrical.

Even at low and moderate speeds, the pitch torque is sufficient to flip the puck many rotations per second. This happens when a shot has misfired. A flipping puck has no lift and increased drag, so the trajectory is not pretty. Like the knuckleball for batters, it can catch the goalie by surprise.

But if the puck has spin, the pitch moment has little or no effect. It won't flutter. This is part of the reason that puck spin is important for stability. There is an interaction between the pitch torque and the spin that is rather complex and goes beyond the scope of this book. Let's just say that the combined effect of spin and pitch torque is this: instead of flipping backward, the puck twists as it travels forward, like the rolling of an airplane when it turns. But this twist rotation is slow, something like one turn per second or so. The twisting is hardly noticeable because the puck would make just a fraction of a turn before it reaches its destination. Interestingly, the puck twists in different directions whether the shooter is left- or right-handed. From a goalie's point of view, the oncoming puck would twist counter-clockwise if the shooter is left-handed, and clockwise if he's right-handed. That is, if the goalie sees this effect at all!

How Far Would a Puck Go?

When a player is caught deep in his own zone and dumps the puck all the way to the other end of the rink, some 200 feet away, one hardly notices the puck slowing down at all. This naturally raises the question: how far would the puck slide, were it not confined by the boards of the rink, and if the ice was very large?

Without any friction, a sliding puck (or any object) would keep on moving in a straight line at the same speed, forever. Of course, ice does cause friction, albeit a small amount of it. Most surfaces offer friction forces that are about half of a body's weight, or a friction coefficient of 0.5, but on ice, it is as low as 0.003. This coefficient depends on variables such as ice temperature, ice roughness, and, perhaps more important, the presence or absence of ice chips (snow). Plugging typical values of ice friction and air drag coefficients into a computer program, we can calculate the distance traveled when the puck is shot at different speeds. (If you need a refresher on the topic of friction coefficients and ice, check back in chapter 1.)

Table 3.3 gives the results of the travel distance for initial puck speeds from 50 mph (a wrist shot) up to 100 mph (a hard slap shot). The first calculations (second column) only take into account ice friction and give distances that are surprisingly large. Putting air drag in the model, which becomes very important at high speeds, the travel distances are reduced by up to a factor of 4. For table 3.3, I used a realistic ice friction coefficient of 0.05, an air drag coefficient of $C_d = 0.54$, and vertical suction forces obtained from numerical simulations.

Although these numbers are estimates and actual results would depend on atmospheric and ice conditions, the general conclusions are the same: to play hockey without boards and still be able to protect the audience, one would need a huge ice surface—and fans would need binoculars.

We now have all the building blocks we need to play hockey: a rink, skates, sticks, and pucks. In "Third Period," we will see how these all come together for the game of hockey. But first, we will look at what is the best we can expect out of a player, the skills goalies need to guard the net, and what can go wrong during a game with injuries (and fights).

SECOND PERIOD

On the Ice and in the Crease

CHAPTER FOUR

The Best

"A good hockey player plays where the puck is. A great hockey player plays where the puck is going to be." This famous advice from Walter Gretzky to his son encapsulates the essence of one of the most important skills in hockey.[1] Trying to understand this quote, some have wondered how The Great One could know where the puck would go next in a game that is as unpredictable as hockey. Or did Walter want his son to stay in front of the net and wait for a pass and a good chance to score?

In hockey, as in other sports, there is a general aptitude called the "feel for the game," a kind of intelligence for reading a play and guessing where it is developing. For an offensive player like Gretzky, this ability would be to know where to go on the ice to get the best chance of creating an imbalance, a shift in the game that would lead to an attacking opportunity. Reading and guessing the play is a continual process that goes on as long as the player is on the ice. In his games with the Edmonton Oilers, Gretzky must have had thousands of thoughts like "If I skate over there and my teammate Jari Kurri wins his battle for the puck along the board, he could pass it to me and we could have a two-against-one attack." In the mind of a legendary player like Gretzky, this kind of thinking would take place in a split-second and be constantly updated as the play evolves.

The feel for the game is an ability that is hard to quantify—you can't attach a number to it—and it is not something that can be explained on a whiteboard because a play is always dynamic: the best place to be on the ice is a function of not only where other players are at a given moment but also where they are going and what they are doing. So it's not something you can acquire from pre-learned "systems of play" because each situation is different—there are too

1. See http://gretzkyhockeyschool.com/pages/walter-gretzky.

many factors to take into account to be summarized by a few basic rules. Instead, it takes a great deal of practice and playing experience to develop the hockey sense that is necessary to score 212 points in one NHL season like Gretzky did. At the same time, I believe this is one of the most neglected skills when it comes to coaching young players, and this void could, in my opinion, be filled by technology and computer simulations. For example, on videos of real hockey games stopped at certain crucial moments during play, young players could practice the art of quickly spotting the areas that seem the most promising for an attacking player or a defensive player. By explaining their choices, they could be corrected and instructed by coaches when need be.

Hockey players develop many other skills that we can analyze in great detail with measurements and numbers. Take fitness, training regimens, and skating and shooting abilities, for example. This chapter examines some of the key features and abilities of the best players in the world.

Playing Hockey, One Explosive Half-Minute at a Time

Long gone are the days when professional hockey players, like Stan Mikita of the Chicago Black Hawks, could be seen puffing a cigarette after a game. Montreal superstar Guy Lafleur used to smoke a pack of cigarettes a day at one point, an amazing habit considering the aerobic and cardiovascular demands of being an offensive player. Nowadays, the level of competitiveness at the top of the hockey pyramid is such that no one can afford to slack off anymore. A multiyear contract is not a free ticket to play just for fun. I remember an instance in 2005 when an NHL star player showed up at training camp in a physical shape that was deemed inadequate—he failed his physical exam—and was suspended by the administration until he returned, slimmed down.

Supervised by teams of trainers, hockey players now follow strict fitness and nutrition regimens to stay in top condition. Training, nutrition, and mental preparation may well be the most important factors responsible for the increased intensity of play at the elite level. It has been shown that the average active man, ages 19 to 30, should consume about 3,000 calories per day to maintain a healthy body mass. By comparison, today's hockey players can burn 1,000 calories in a single game.

In terms of athletic demand, playing hockey is characterized by short bursts of intense activity alternating with a few minutes of rest. A typical offensive player goes on the ice about 10 times in one period of 20 minutes, each shift lasting for about 30 seconds (sometimes longer when a line change is not possible). Some players log in more time than others, but with a total of about 15 minutes of play, 75 percent of game time is spent recuperating on the bench. The team's top defensemen take more shifts and stay on the ice for a total of about 20 to 25 minutes.

These play times are small compared with those of some other sports. For example, it's not rare for basketball players to stay on the court for 35 minutes or more. In soccer, many players stay on the field the whole 90-minute game. By comparison, hockey is by and large an anaerobic activity and more like sprinting a 200 meter dash every few minutes or so.

The difference between aerobic and anaerobic activities is in the way energy is extracted from glucose, the body's source of energy. The aerobic energy system requires a flow of oxygen to the muscles via the bloodstream, and this oxygen is used by the cells along with glucose to create adenosine triphosphate (ATP) to power the muscles. ATP is what the muscle cells run on. After just a few minutes of physical exercise like running, swimming, or skating, the aerobic system already provides most of the energy to the muscles, and so the rate of oxygen intake by the athlete's lungs increases with the power he is delivering during this effort.

The anaerobic system, in contrast, relies not on a steady flow of oxygen but on the energy stored within the muscles, ready for immediate consumption. This pre-packaged energy source includes ATP and glucose. Anaerobic mechanisms of energy extraction are more powerful, but they have a few drawbacks: they are less efficient and they are short-lived. The inefficiency is from the glucose being only partially broken down—we can compare the process to pumping gas into a leaky fuel line; the engine gets a boost but some of the fuel is wasted. The short life of anaerobic energy extraction is the result of the finite amount of ATP already present in muscle (which is consumed in about 10 seconds of intense activity), and from the lactic acid that is created when glucose is consumed (metabolized) without oxygen. This acidification produces muscle fatigue and a burning sensation. As the activity goes on, the aerobic energy system gradually takes over (on a time scale of 2 to 3 minutes), and the muscles are provided with a steady supply of ATP, albeit at a lower, more sustainable rate. But this takeover is never complete and never completely shuts down the anaerobic systems: during an aerobic activity, they can be resorted to whenever an outburst of effort is needed, like during the final sprint of a long-distance race.

The chances of an athlete reaching the top levels of hockey and staying there are closely linked to his fitness, so it is not surprising that teams monitor and measure athletes regularly. Every year, before the NHL Entry Draft, many top prospects gather to showcase their physical abilities in front of NHL team managers. A battery of tests is performed: body composition, aerobic and anaerobic capacity, eye-hand coordination, strength, power, endurance, and flexibility. The aerobic and anaerobic tests are some of the most revealing in terms of an athlete's potential.

A common method for measuring anaerobic fitness is the Wingate VO_2 test. An athlete sits on a stationary bike and pedals as hard as he can against a resistance that represents a fixed proportion of his body mass. Athletes typically attain power output averages of 10.1 watts (W) per kilogram (kg) of body mass, and some of the best will reach the 15 W/kg level.

Aerobic fitness is assessed by measuring the maximum amount of oxygen uptake by the athlete while he is pedaling on a bicycle ergometer or running on a treadmill. The pace of activity is slowly raised until the maximum volume of oxygen per minute is extracted from the bloodstream of an athlete, a quantity called the "VO_2 max." This measures the maximum cardiovascular output of the athlete. The best results for hockey players are typically in the mid-60 milliliters per kilogram per minute (ml/kg/min) and the average around 55 ml/kg/min. By comparison, a performance by a typical young adult male would be around 40 ml/kg/min.

Over the years, results of aerobic and anaerobic tests in hockey have shown an upward trend. This was measured in a study on the performance profiles of players from one NHL team from 1979 to 2005,[2] showing that the sport is increasingly competitive and fast paced, requiring much more from its players than it did from the (cigarette-smoking) stars in times past. Any doubt about the rise of pace in hockey can be removed by comparing footage from the 1976–77 Montreal Canadiens (they had an incredible 60-8-12 record) with that of any current NHL team: the average player today is faster, stronger, and more skilled and plays more intensely than the stars of decades ago.

Fastest Skaters

Hockey is an exciting sport to watch partly because of its fast pace. Speeds near 35 km/h (22 mph) have been measured when players race to the puck (as in figure 4.1), and this is getting close to Olympic sprinter Usain Bolt's top speed of 44 km/h (28 mph). Higher speeds would be reached if players had more space to accelerate or did not have to turn around all the time. Yet, in spite of this blistering pace, it is amazing that some players manage to stand out, making it look easy to break away from other players chasing them. Videos are available online showing players like Michael Grabner, Sidney Crosby, Erik Karlsson, and Steven Stamkos (seen in figure 4.1) doing just that.

For many years, the NHL held the fastest skater competition in a format where fully equipped players would skate around the rink (and behind both nets) as quickly as possible. The winner usually made it in about 14 seconds. The record holder is Mike Gartner, who clocked 13.4 seconds in 1996 at Boston's Fleet Center, which, unlike the smaller Boston Garden, is a modern, standard-size rink. But back then, the lap was a little shorter because the nets were pulled closer together, placed in front of the goal line to leave more space for the skater to go around. This shortened the lap by approximately 5 meters, and taking this into account would put Gartner's time at around 13.8 seconds in a normal lap; this is still a very good time, but probably not as fast as today's best skaters. Still, Gartner was a natural skater known for his stride power and efficient crossovers during turning and he was

2. H. A. Quinney, R. Dewart, A. Game, G. Snydmiller, D. Warburton, and G. Bell, "A 26 Year Physiological Description of a National Hockey League Team," *Applied Physiology, Nutrition, and Metabolism* 33, no. 4 (2008): 753–60.

FIGURE 4.1. In a race for the puck or during a breakaway like this one by Steven Stamkos, powerful strides and a good skating technique are essential. (AP Photo/Mike Carlson)

36 years old when he set his record, almost a decade past his prime. Backward skating competitions have also been held for defensemen, who complete a lap in about 16.5 seconds, faster than many of us could do while skating forward.

The approximate length of a lap around the 2 nets is 160 meters. With a lap time of 14 seconds, the average speed is 41 km/h (26 mph). But that's not the top speed players can reach since they start from rest and it takes time to reach maximum speed. The second half of the lap is done at full speed in about 6.5 seconds, which matches Usain Bolt's running speed record.

In 2008, the NHL All-Star weekend held the competition in a linear race. Starting at rest at the goal line, players raced to the far blue line, some 34 meters away. The best time was 4.7 seconds, giving a top speed exceeding 40 km/h (25 mph).

It is interesting to compare the skating performances of hockey players with that of speed skaters. Watching short track speed skaters go around the rink at the Winter Olympics, it looks like they are going around a small circle. In fact, the so-called short track ice is nothing less than a regular-sized hockey rink. The loop perimeter is 111 meters (the minimum distance to cover in 1 lap) and speed skaters go around it in less than 9 seconds. They attain top speeds of 50 km/h (30 mph) or more. Of course, speed skaters are specialists, and specialists are the very best at what they do. Hockey players have to master another set of skills in addition to their skating ability. The same is true when comparing the jumping abilities of basketball players with that of high jumpers in track and field competitions. A slam dunk is always impressive, but the vertical leap basketball players reach in doing so would not get them even close to a medal at high jump competitions. At some top-level events, the bar is set at 2.4 meters above the ground, and jumpers have to raise their center of gravity very close to that height to clear the bar. If they were to jump and keep their body up straight (like basketball players do) instead of arching their bodies, their heads would rise higher than the rim so they could peer down the basket. Their vertical leap is around 1.3 meters and their heads would clear the rim by about 65 centimeters. Meanwhile, when basketball players jump to the net, the tops of their heads do not quite reach the same level as the rim. But again, high jumpers are specialists.

Hardest Shooters

The Hardest Shot competition is to the NHL's All-Star weekend what the 100 meter sprint is to the Summer Olympics. It is a marquee event. A puck is put in front of the net some 9 meters (30 feet) away, and players, starting no farther than the nearest blue line, then skate toward the puck and slap it to the net as hard as possible. The puck, initially at rest, reaches the goal in two-tenths of a second, roughly the same time as the human reaction limit. The puck speed is measured by a radar gun and projected onto a screen for fans to see. Fans cheer the loudest when they see the 100 mph mark broken. The NHL's current record is held by defenseman Zdeno Chara (seen in figure 4.2) at 108.8 mph. Shea Weber, another defenseman known for his powerful shot, came very close to taking the title in 2015 when he shot at 108.5 mph.

Slap shots reaching speeds of 100 mph are rare in actual hockey games, for two reasons. First, in a game, players don't take a few strides toward a puck and slam it; they carry the puck next to them and hit it while they are on the move. In the Hardest Shot competition, the speed of a player helps transfer more momentum to the puck.

FIGURE 4.2. Zdeno Chara shoots a puck at 108.8 mph to win the NHL's Hardest Shot competition in 2012. (AP Photo/The Canadian Press, Fred Chartrand)

Second, the 100 mph barrier is broken by the best NHL shooters only about half the time under the best of circumstances. Table 4.1 lists the puck speeds measured at the 2012 Hardest Shot competition and shows that even the best shooters don't reach that mark each time. In an NHL game, a typical slap shot speed would be in the 85- to 95-mph range.

The radar gun placed behind the net measures the puck speed based on how quickly it comes toward it. In that configuration, a puck moving vertically registers a speed of zero. This is why police officers can measure speed when they are in front or behind a vehicle, but not when they are beside it. So when a puck is shot high toward the net instead of along the ice, the vertical component of its velocity is not registered. The measured speed will be lower than it is in reality. But how much difference does it make? This can be estimated using mathematics, and comparing two shots, one made at ice level (directly toward the radar) and one toward the center of the net, the difference is about 0.2 percent. According to table 4.1, that difference is not

Table 4.1. Puck speeds at the 2012 All-Star Hardest Shot competition

Player (attempt)	Speed (mph)	Player (attempt)	Speed (mph)
Chara (2)	108.8	Spezza (1)	100.4
Chara (4)	107.0	Adam (2)	98.3
Chara (3)	106.9	Phaneuf (2)	97.9
Chara (1)	106.2	Phaneuf (1)	96.9
Weber (4)	106.0	Adam (1)	96.4
Weber (1)	104.9	Faulk (1)	95.9
Weber (3)	102.6	Wideman (2)	95.3
Alfredsson (2)	101.3	Wideman (1)	94.6
Alfredsson (1)	101.1	Faulk (2)	93.5
Spezza (2)	100.5		

Note: Average speed = 100.8 mph.

likely to be a deciding factor between first and second place at the Hardest Shot competition. To register the highest possible speed, there's perhaps another reason for trying to keep the puck low (without touching the ice), and it has to do with the tendency of a projectile to slow down when it goes up. But again, using the previous example, the difference is only about 0.3 percent.

As it turns out, air resistance is a far more important factor. We saw in chapter 3 that air drag on a puck is significant at high speeds. We know that a puck loses 1 percent of its speed per 10 feet of traveling due to air drag. If the radar measures speed when the puck reaches the net, a 100 mph shot could have lost 3 mph by then. But at the competition, since the same effects of aerodynamics apply to all shooters, air drag is not likely to be a deciding factor.

On the other hand, it is legitimate to question the accuracy of the radar gun technique. A number like 108.8 mph implies an accuracy of four significant digits. In the jargon of scientists, significant digits represent results of meaningful measurements. For example, if the radar is only precise to the nearest 1 mph, then the 0.8 mph figure is meaningless and is considered nothing more than random "noise." Given that the puck loses speed at the rate of 1 percent per 10 feet

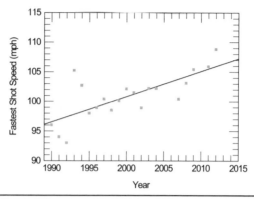

FIGURE 4.3. Highest speed recorded at the NHL Hardest Shot competition since it began in 1990.

because of air drag, if the puck speed is measured over several feet, the measurement uncertainty could be on the order of 1 mph. It would then make sense to round the result up or down to the nearest mph.

Figure 4.3 shows the increasing speed of the fastest shot over the years. The trend suggests that the 110 mph mark will be reached in the near future. Let's not forget, however, that these data are only for NHL players. It is likely that the fastest shooter in the world is not an NHLer, just like the longest golf drivers do not play in the top-rated league (the PGA Tour). In fact, in 2011, Denis Kulyash of Avangard Omsk set the bar at 110.3 mph at the annual Kontinental Hockey League (KHL) All-Star Game. The KHL is Russia's top professional league. Then in 2012, Traktor Chelyabinsk's defenseman Alexander Ryazantsev made a shot that was clocked at 114.1 mph. While again there may be reasons to question the radar accuracy, it should be noted that the KHL's competition puts the puck a little closer to the net and allows the player to gain speed from farther away, two factors that may contribute to faster speeds.

Playing Performance and Age

Different factors contribute to changing a player's performance over time. Beyond a certain age, the body's regeneration mechanisms become less efficient, secreting less human growth hormones for instance, so it takes longer for the athlete to recover from an intense game and longer for injuries to heal. Past the age of 25, there is also

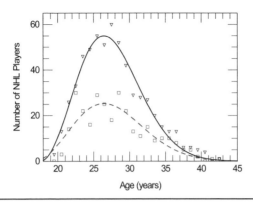

FIGURE 4.4. NHL players' age distribution for forwards (triangles) and defensemen (squares) during the 2011–12 season. The curves are added as guides to the eye.

a tendency for the body to slowly lose muscle mass, thereby delivering less power and speed. As a result, only 1 percent of NHL hockey players celebrate their 40th birthday while still active in the league. Figure 4.4 shows the age distribution of all NHL players. The curves are guides to the eye and appear as slightly skewed bell curves. The distribution of forwards and defensemen peaks at around 26 or 27 years of age. As we can see, the popular belief that defensemen stay in the league longer than forwards is not true. The age distribution of hockey players has stayed relatively constant through time and is also similar to that of other sports, like soccer. For example, at the FIFA World Cups between 2006 and 2014, the average age was 27.4 years.

The distribution curve represents the number of players in each age category, but it also shows the likelihood that a player will be active in the NHL at any given age. Athletes are most likely to play at the elite level between the ages of 21 and 32. This is a rough indication of how hockey players progress and recede with age. If there are more 27-year-old players than any other age, it is because athletes tend to be at their best at around that age. Of course, this number is a statistical average only: we know plenty of players who peaked later or earlier.

Notice how the age distributions in figure 4.4 are not symmetrical, sloping more gently on the side of older players. For a player, this means the likelihood that a young player will be in the NHL increases rapidly with age until 27 and then slowly decreases afterward.

Age distribution is not an exact representation of hockey fitness. For instance, we should not expect the average 27-year-old hockey player to be twice as good as the average 21- or 32-year-old player because there are twice as many of them. But we will see later how we can estimate hockey fitness with age from the age distribution curve.

Here's an interesting question: are older NHL hockey players less productive than the ones near their peak? Looking at figure 4.4, we might indeed expect players near the 27-year peak to perform the best. So let's look at player productivity measured by the number of points per game. Figure 4.5 plots productivity with age. Lo and behold, productivity is practically flat with age, with a slight upward trend. What is going on? It looks like veterans stay in the NHL not just for the fun of it; they are just as competitive as younger players. We have some kind of paradox: hockey players appear to become better with age, yet veterans quit in large numbers.

To explain this puzzle, we need to be aware that figure 4.5 only tells one side of the story: it only counts the older players who stay in the NHL. Older players who are able to stay and compete also tend to represent the cream of the crop. Many of them are former stars. While they are past their peak, many are still more productive than the average player. This was the case with Wayne Gretzky, who was still an above-average hockey player when he retired at the age of 38.

It is generally believed that veterans play fewer games in a season because they are more prone to becoming injured. Yet that is not necessarily the case either, as figure 4.6 shows. In fact, the average number of games played increases with age. An explanation for this might again be the filtration effect: those older players who stay in the NHL tend to be the most talented ones, and they may also train more and keep themselves in better physical shape. Teemu Selanne, who played 82 games in 2011–12 at 42 years of age, is a good example. This is all the more remarkable because Selanne is a forward, a position that demands speed and agility. Another reason that older players play more games might be because coaches appreciate their leadership and consistency, in other words.

So, we have seen players who have an incredible energy, players who skate at top speeds, players who shoot amazingly fast, and players who perform at levels we amateurs can only dream about. But how

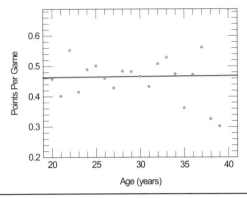

FIGURE 4.5. Average points per game by forwards who played more than 20 games during the 2011–12 NHL season. The line is a linear best fit.

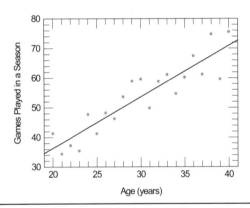

FIGURE 4.6. Average number of games played by NHL forwards in the 2011–12 season

long can professional hockey players sustain these impressive skills before they have to retire?

How Much Does a Player's Fitness Change with Age?

We will now try to estimate how fitness peaks and decreases over time for hockey players. For each age group in a population, there is a fraction of individuals who can make the cut and compete in the NHL, but that fraction does not remain the same with age. Figure 4.7 illustrates the idea. Three age groups are compared in terms of the

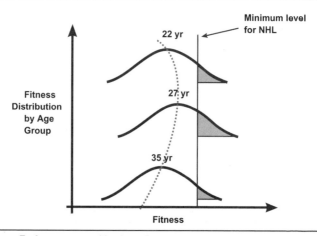

FIGURE 4.7. Each age group of hockey players has a different distribution of fitness, notably in terms of the group's average, represented by the dotted line. The gray portions represent players who meet the minimum fitness level necessary to play in the NHL.

way fitness is distributed among players of the same age. Most notably, it is a group's average fitness that changes with age. The 27-year-olds are shifted to the right; in other words, they have higher fitness levels compared with the 24- and 35-year-olds. On average they are better, so more of them make the cut (meet the minimum level for the NHL). But each age group has some players who can make the cut too, and these players are represented by the areas shaded in gray. The dotted curve represents how average hockey fitness changes with age, and this is what we would like to estimate.

To proceed with this model, we need two additional pieces of information. First, we need to know how hockey fitness is distributed within a certain age group, in particular the standard deviation, or the spread. As it turns out, when measuring various skills and features in a group of individuals, generally the distribution looks like a bell curve, and most individuals fall within ±15 percent of the average. Many human traits and skills follow this pattern, including height, academic scores, running speed, and so on. So we will assume that it is also the case for hockey fitness, whatever way we would choose to measure it. If an age group has a fitness distribution centered at 100, we will take a bell curve centered at 100 with 15 as the standard deviation.

The second piece of information we need is the minimum fitness level required of an NHL-caliber player. This can be estimated based on the percentage of all hockey players who make it to the NHL. As of 2006, 1 out of every 9,700 Canadian 27-year-old men played in the NHL. Of course, not every Canadian male has played or has wanted to play hockey. But the sport is very popular, and we can estimate that about half of Canadian men have had a chance to play organized hockey at some point during their childhood, so at the very least that they would have had an opportunity to test their skills and decide whether or not to pursue the sport seriously. For that subgroup, the odds of making it to the NHL are about 1 in 5,000. Assuming a bell-curve distribution in fitness, these odds imply (using statistical methods beyond the scope of this book) that the NHL player must be at least 3.6 standard deviations above the mean. In other words, if we peg the average skill (fitness) level of 27-year-olds at 100, the minimum level to be an NHLer would be 155.

Based on this model, we now find how each age group must be positioned in figure 4.7 to account for the age distribution of figure 4.4. The results are shown in figure 4.8. Notice how between ages 21 and 35, players are still within 5 percent of their peaks. Also, younger players gain fitness faster than older players lose fitness during their decline, which enables athletes to have longer careers. The typical 40-year-old should still be at 91 percent of his peak ability—provided he kept training, of course. Interestingly, if we made two hockey teams, one with the best 18-year-olds and one with the best 42-year-olds, it appears that it would be a fairly even match.

Figure 4.8 only represents the average loss of hockey fitness beyond the peak age. Of course, each athlete has his own progression and decline, and some peak later in life. As they get older, some of them adhere to strict training and diet programs that compensate for the natural decline of the body. Adjustments can be as simple as drinking plenty of water regularly to compensate for the body's diminishing ability to hydrate itself.

Other factors may limit the playing career of some players, such as the following:

- Injuries, which are different from physiological aging per se, can end a career prematurely. Each playing season brings chances of

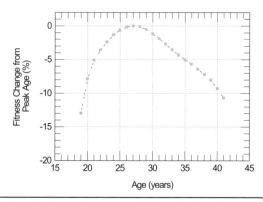

FIGURE 4.8. Estimated changes in playing ability in hockey players over time: the average fitness is lower on either side of the typical peak age of 27.

serious injury. Therefore, the odds of a player's having suffered a debilitating injury, like a severe concussion, by the age of 35 are much greater than that of a 20-year-old. For that reason alone, even if player fitness did not change over time, there would still be fewer older players than younger ones.

- Success can decrease drive and motivation. Older players have become economically secure and often have to tend to families, and their popularity has opened up alternative careers outside playing hockey, such as coaching, management, and scouting. Some players, most notably European ones, also choose to move to less demanding professional leagues, sometimes closer to home.

- High expectations bring criticism with aging. Veterans are often compared against themselves in their best days. Some superstars, like Wayne Gretzky and Mario Lemieux, could probably have competed for many more years, but they might not have been as satisfied with just being average. For example, when 35-year-old Mats Sundin produced 1.1 points per game in 2005–6 (twice as much as the average NHL player), fans who seemed disappointed would often say how Sundin was past his prime.

- Some specialty players like enforcers, checking defenders, and penalty killers may find it difficult to keep up with the competition and may retire early. Not all enforcers are still willing to drop their gloves and fight for their team's honor in their thirties.

Other than keeping himself in the best possible shape, an aging athlete could prolong his career by playing fewer games. Older players may not be able or willing to play 82 games in a season, but they may well compete in 40 games productively. Indeed, some players have come out of semi-retirement and joined teams late in the season with great success. But to accommodate such a "part-time" playing arrangement, teams would have to adjust their roster continually. It would also potentially take away jobs from young players who would otherwise play full-time.

A Proposal for a Different All-Star Game Format

The NHL All-Star Game has been held almost every year since 1947, and it has taken different formats. It has pitted the Stanley Cup defending champions against the rest of the league, the NHL All-Stars versus the Soviet Union, one conference against another, the East against the West, North America against the World, and more recently, Team X against Team Y, with X and Y being two well-known players acting as captains who select their own roster from a pool of players.

Each one of these All-Star formats involves some form of friendly rivalry that is necessary to make the game interesting to watch. Something is at stake. Athletes want to display skills and prove themselves. Still, many fans don't watch the All-Star Game because, according to them, it lacks the seriousness and intensity of a real hockey game.

The All-Star teams have never been divided according to age, but I think it would be an interesting and revealing experiment. It would bring a new form of rivalry in which each player would have something to prove. On one side, veterans would want to show the fans that they are still very much on top of their game; on the other side, young players would want to prove that they, the rising stars, are better. The styles of play could be contrasted as well, pitting experience against raw speed and talent, and perhaps defensive play against offensive play.

From the perspective of the fans, it could give them a reason to cheer for one team or the other based on their own age group. At first, I'm sure many fans would predict a sweeping victory for the younger team, but according to what we've discussed in the previous section, the competition could very well be evenly matched. In fact, I would predict a slight advantage for the veterans . . . and I'm not biased!

The Science (and Art) of the Scout

I like hockey pool magazines. Every late summer, hockey publishers produce special issues dedicated to the upcoming NHL season. Nice, glossy pages with colorful tables highlight the predictions of hockey experts (I suppose) about players and teams. They predict assists, points, and, in the case of goalies, games won. They even boldly predict how many games each player will miss in the season, taking into account the odds of injuries. Hockey fans use these complete reports to help them decide on whom to select for their hockey pool.

The only problem with hockey predictions is that they are often wrong. I suppose many hockey fans don't realize this because they discard their old magazines and never bother to compare predictions and facts. After suspecting for years these magazines were not as helpful as I thought, I once kept my magazine until the end of the season to compare results. (The magazine was from a reputable publisher that I won't mention because I assume they are no better or worse than other publishers.) The end results were astonishing, bordering on humorous. I looked up the top 50 forwards and compared their rankings as predicted by experts with their actual ranks at the end of the season. The experts' predictions were off by 43 ranks, on average. The predicted productivity was off by 19 points, on average, which represents almost 30 percent of the average point production.

As it turns out, it is very difficult to predict the performance of athletes in an upcoming season. And this is true even though we know these athletes well, we know their past NHL performance and all previous statistics, we know what teams they will be on and which line mates they are likely to have, and so on. Now imagine if we were facing the task of predicting the long-term performance of newcomers, those athletes who have never been tested in the NHL and are not yet fully developed, whose line mates and teams we don't know yet. How hard would that be? Yet this is the nearly impossible task facing the scouts who are trying to find the next generation of professional hockey players.

Given the quantity of unknowns surrounding new players, it is perhaps not surprising that the accuracy of scouts and NHL teams in selecting new talents for their rosters is not much better than that of hockey pool magazines, in spite of the fact that some of these draft picks are long-term, multimillion-dollar investments.

Let's look at the numbers. Each year, during the NHL Entry Draft, more than 200 new players are added to the league in 7 rounds of drafting. The first-round picks are the most coveted ones: they are supposed to be the 30 most promising hockey talents. Yet, out of these top 30 hockey prospects, only half will get to play 400 games in the NHL, a relatively modest number used as a benchmark of success, or the equivalent of about 5 seasons played, and a bit short of an average NHL career length. Certainly there is space for more than 15 new players to be added each year in the NHL. In fact, given the career length of a typical player and the number and roster size of NHL teams, about 100 new players could be added each year, of which at least 50, we could safely say, should have an average or above-average professional hockey career. (For those who want to do the calculations along with me: we assume 30 teams with a roster of 20 athletes each, totaling 600 players in the NHL. With an average career length of 5.7 seasons, an average of 3.5 players per team drop out of the NHL each year, which corresponds to 105 places to be filled by new players.)

So each time an NHL team picks a first-round draft, chances are roughly 50 percent that it will be a miscalculation. This reality has two implications: (1) there is a high proportion of early draft picks that are wasted, so to speak, and (2) it's possible to compete as a team without early picks because some good talent is always available in later rounds. For example, the Detroit Red Wings managed to remain a powerhouse in the NHL without having a single top-10 draft pick for a very long time. Their highest draft pick of the past 2 decades was in 2005, and he ranked 19th overall: defenseman Jakub Kindl. Yet they dug deep and found gold nuggets in future stars like Pavel Datsyuk, selected 171st overall in 1998, and Henrik Zetterberg, selected 210th overall in 1999. These were steals, so to speak. But before we wonder what kind of crystal ball the Red Wings administration was using, let's be reminded that they also picked six players before Datsyuk, in the same entry draft, who produced a total of zero points in the NHL. In the case of Zetterberg, 3 players were drafted by Detroit before him, 2 of whom never played a single game in the NHL and the other played a total of 32. The potential of both future stars was vastly underestimated, and the Red Wings administration had plenty of chances to pick them but miscalculated and bet on others. On the other hand, they get full credit for picking Nicklas Lidstrom 53rd overall in 1989, when everyone else had this now-legendary defenseman ranked much lower on their lists.

Then there are other extremes, like Martin St. Louis, one of the NHL's top scorers, who was not even technically drafted. He had to produce incredible numbers year after year at the college level before an NHL team took notice and signed him up (it was the Calgary Flames). Signing undrafted players is more common nowadays. It makes sense, as some players take longer to develop.

Of course, no hockey scout would realistically expect to pick a future star at each and every entry draft pick. Inevitably there will be players picked who range from the middle ranks to below average, and teams need these players too. When we think of an NHL team, say the New York Islanders, we often think of the 20 or so men who make the roster. But in a way, the roster runs much deeper than this: it includes the farm teams, which (for the Islanders) are currently the Bridgeport Sound Tigers, operating in the American Hockey League, and its mid-level professional East Coast Hockey League affiliate, the Stockton Thunder. The separating line between the parent team and its affiliates is a bit fuzzy. Farm team players are regularly called up to play in the NHL to replace injured players or players who are not performing well. In a way, roster-depth is a kind of safety net for the parent team.

Some farm team players are actually good enough to have a permanent position on an NHL team but they don't, either because there is no space on the parent team at the time or they are considered too young and the parent team wants to develop them further. A good example is Jason Spezza, who was a second overall NHL pick in 2001 but had to wait until 2005 to permanently join the Ottawa Senators. In the meantime, he played in the farm system, starting with the Windsor Spitfires of the Ontario Hockey League.

But for the majority of farm team players, the dream of having a long, successful career in the NHL never materializes. I imagine it must be difficult to be in such a perpetual position of "almost making it" to the big league. After some time, many quit to join leagues overseas where their talent is perhaps more appreciated and salaries are still reasonable.

Birthdays Matter

We now look at the surprising phenomenon of how a player's birthday influences his chances of playing professional hockey. It's a little-known fact that the NHL consistently drafts more players who are born in the first quarter of the year (January to March) than those born in the last quarter (October to December).[3] And the ratio between the two groups is not small: it's 2.5 to 1. This has been attributed to a bias starting at a young age, in minor hockey leagues, where the cut-off birthdate is set at December 31. So children who are born early in the year find themselves in groups where they are the oldest, and thus they tend to compare favorably in terms of size, skill, and maturity. The small advantage results in a selection bias and more opportunity to play on better teams and to develop as players. The results are especially apparent in junior leagues. For example, out of the pool of players drafted into the Quebec Major Junior Hockey League in 2013, the ratio between players born in the first and fourth quarters of the year was 3.3 to 1 (and 2.3 to 1 between the first and second half of the year). The NHL, in contrast to minor hockey leagues, has a cut-off birthdate set at September 15 respective to the year's entry draft for a team to sign an 18-year-old. This would technically favor players born in the last quarter, but because the pool of top junior players, from Canada especially, is already loaded with players born in the first quarter, it makes no difference.

There is a somewhat counterintuitive consequence of this: those players who are born later in the year and who are nonetheless drafted by the NHL tend to exhibit superior talent, and indeed statistics show that they perform better than average in the league. A greater percentage of them reach the list of top scorers. For example, of the top 50 Canadian NHL players in February 2014, those born in the second half of the year accounted for almost half of this elite group even though there were fewer of them in the league. Meanwhile, middle-of-the-pack and bottom-of-the-pack NHL players are overly

3. R. O. Deaner, A. Lowen, and S. Cobley, "Born at the Wrong Time: Selection Bias in the NHL," *PLOS ONE* 8, no. 2 (2013).

represented by players born early in the year. Currently, Canadian NHL forwards born in the last quarter are 57 percent more likely to be top scorers (producing more than 0.5 point per game) than the same group born in the first quarter, and they are 45 percent less likely than first-quarter–born players to be average to low scorers (less than 0.5 point per game). The same bias is also there, although less important, among non-Canadian players: the respective figures are 27 percent and 22 percent.

It appears as though the birthday bias does not impact negatively players whose skills are at the very top, but it does make a difference for players who could be journeymen in the NHL and lower leagues. As a result, first-quarter–born players fill the ranks of farm teams and elite junior leagues. To take the previous example, on the current roster of the Windsor Spitfires, those born in the first half of the year outnumber those born in the second half 27 to 4. On the Bridgeport Sound Tigers roster, it is 25 to 18.

The consequence of this bias, of course, is that some talent is being wasted. If a hockey nation could tweak its hockey system to find a way to tap into this talent, it could provide a considerable advantage. Like many other abilities, hockey-playing abilities are a combination of talent and practice. Without one or the other, the player doesn't go very far. The ideal combination of talent and practice is still an open question, but what we know for sure is that the current age bias selects in favor of individuals who have had more practice (and more time to grow), and this amounts to favoring practice over talent.

A possible solution would be to make space for a greater variability in the way young players are ranked in various divisions. In minor hockey leagues, coaches in the highest division pick the best players, and then the second division coaches pick the best among the rest, and so on. When players are selected based on the skill level coaches see, this automatically creates an age bias. Taking into account the amount of practice a child has had would perhaps level off the field. For example, for socioeconomic reasons, not all children have the opportunity to play in summer leagues and participate in pre-season training camps. These activities require parents with extra time and money. At the start of the following season, children who could do them tend to shine over others who spent the summer doing other things but may have more potential. Like polo, hockey has been dubbed a "rich man's sport" by some.

Another good reason to incorporate more variability is the fact that each person develops physically and mentally at his own pace. We can see that when comparing children in the same school class. Height varies greatly, especially at a young age, and a child's height at a given age is not always indicative of how tall he will be as an adult. The same applies to mental development, and educators have known this for a while too.[4] The pace of physical and athletic development also varies greatly from one person to another. But hockey development and drafting systems don't take account of this variability. In a way, it is surprising that almost all newly drafted NHL players are 18 years of age because it will be almost a full decade before they peak. Most of them play their first game only by the time they reach 22. Yet the average career length in the NHL is just over 5 years long, which means that in an ideal world, the average player should start playing at around 24 and quit at around 29, thereby playing an equal amount of time on each side of his physical peak. But the average age of retirement is 28 years,[5] and many players retire before they reach their peak. By drafting and consistently investing in 18-year-old players, NHL teams are selecting in favor of players with precocious development but are not necessarily getting the best talent.

Similarly, NHL teams may well benefit from a system in which their own drafted players are given more equal chances to compete and develop. This may be a bit harder for them, however, as players who sign expensive contracts are, in essence, investments by the team. Higher draft picks, for example, tend to be given more playing time on the parent team (and less on the farm team), which inevitably comes with a better environment and more resources to develop. This "locking-up" effect ensures that they will produce more than the other players relegated to the minor leagues, but in the end it may not have been the best thing for the team. In a way, deciding on the future of an athlete at 18 years of age is premature given that physical development is not uniform among individuals.

---

4. See J. Medina, *Brain Rules: 12 Principles for Surviving and Thinking at Work, Home, and School* (Seattle: Pear Press, 2008), chapter 3.

5. See www.quanthockey.com/Distributions/RetireeAgeDistribution.php.

Of course, to change a hockey development system would never be easy in the best of situations, and as is often the case, the pervasive element of personal influence and old mindsets may turn out to be the biggest barrier.

We have taken a glimpse at the best of the best and how teams choose and groom the next generation of players. Let's now turn our attention to the player whom many feel has the most critical position on the team: the goalie.

CHAPTER FIVE

The Goalie

Minding the Net All the Time

Some people think that goaltending is just about being quick and flexible, but you can have cat-like reflexes and good eye-hand coordination and still not be able to keep the puck out of the goal. This is because effective goaltending requires a set of complex positional techniques and mental skills that are only acquired through practice and coaching.

The goalie's job is to block shots and either keep the puck (no rebound) or deflect it away safely. To do this, he needs to cover as much area of the net as possible, making the openings as small as possible for the shooter. The goalie must try to "look big" from the shooter's point of view. The key point, as we will see, is that reflexes are just not good enough. There's a limit to human reaction time, something like two-tenths of a second, the implication of which is that when a shot is taken from within a certain radius of the net, all the goalie can do is hope the shot hits him and not the net. The less space of the net left uncovered, the less chance there is for his opponent to score. Figure 5.1A shows a good example, where Cam Ward blocks a shot taken at close range by using a kind of "default" posture that minimizes the opening to the net. In figure 5.1B, Carey Price uses a similar technique, but since the shot is taken from farther away, he has enough time to move his hands and catch the puck with his glove. In the latter case, good reflexes are helpful.

As the players move around, the goalie positions himself in front of the net to cover the angles; each situation calls for a different trick and a different posture. When the shooter wraps the puck around the net, the goalie's posture depends on if the puck arrives from his glove side (the left hands of goalies in figure 5.2) or from his blocker side. On a breakaway, the positioning technique is not the same as

FIGURE 5.1A. (*top*) With too little time to catch the puck, Cam Ward takes a posture called the "blocking butterfly," which minimizes the opening to the net from the shooter's viewpoint, and stops the puck with his elbow. (Jacob Kupferman/CSM, Cal Sport Media via AP Images) **FIGURE 5.1B**. (*bottom*) The butterfly technique is effective against shots taken from far away, but in this case the hands can react to make the save, like Carey Price of the Montreal Canadiens does here. (Icon Sportswire via AP Images)

when there is a defenseman from his team in front of the opponent, which is again different from when two attackers come against one defenseman, and so on. The goalie must be ready to quickly skate frontward, backward, and sideways, as well as go down on his knees, move side to side with one knee down, and get back up again. To add to the difficulty, at all times he must keep his weight on balance on his skates to face the shot and to be able to move efficiently (see

FIGURE 5.2. A stable stance (*top*) is best for making saves, while being off balance is not (*bottom*). (Photo Credits: Nick Doucet and Moncton Wildcats organization)

figure 5.2). For this reason, it is sometimes said that the goalie is the best skater on his team.

When a goalie forgets one of his tricks—and there are so many it's easy to do so—his game suffers. Like the golfer experiencing difficulties who needs to find what he must tweak in his swing, professional goaltenders and their coaches continually monitor for weaknesses and make practice drills to address them. This is why Martin Brodeur, the NHL's all-time leader in regular season wins, was said to prepare for each game by viewing the previous game on video, analyzing each moment he touched the puck, just to reinforce the mechanics and make tweaks if need be.

Outside the technical aspects, there is another side of goaltending that is almost never discussed, and yet it probably accounts for much of the difference between professional goalies and amateurs: the mental aspect. Goaltending is as much mental as it is physical. The goalie also has responsibilities similar to those of the quarterback in football and the pitcher in baseball. Unlike other hockey players, who don't necessarily need to be focused all the time, a goalie can be caught off guard by a momentary distraction. Such distractions can be as simple as wandering thoughts. To be at his best, the goalie needs to keep his mind on the play and his eyes on the puck. It has been measured scientifically that the goaltender has greater chance of making the save when he's observing the puck at the moment the shot is taken.

If a goalie could pick an ideal zone to be in, it would be one where he's mentally focused and physically relaxed, his eyes tracking the puck while his body moves in a quick, fluid fashion. In this ideal mental zone, the "upper" regions of the brain analyze the ever-changing layout of the game and the position of the players, while the preconditioned, or "lizard," part of the brain takes care of reacting to shots and the already-learned mechanics of making a save. Overanalyzing can only interfere with this process. When a goalie is nervous and thinking too much, it often looks as though he is "trying too hard."

Why is it hard to stay focused? Part of the problem is that a goalie is in the game for the full 60 minutes of play (unlike other players, who play no more than 20 minutes) and receives only a total of about 1 minute of real action time. That 1 minute is 30 shots at 2 seconds per shot—1 second each for blocking the shot and controlling the puck. And what is there to do for the remaining 98 percent of the time? Move around the net, keep the angles closed (block as much of the net as possible), keep your head clear, and get ready for the next shot.

Sometimes after a game where I received few shots, teammates would remark how easy the game must have been for me. Ironically, these are some of the more difficult games for a goalie because it's harder to maintain focus. The worst situation for a goalie is a game where he receives little action except for the odd breakaway or point blank shot. At the 2014 Winter Olympics, Latvia came close to upsetting Canada in men's hockey. It was the quarterfinals game. The Canadians dominated, as many had expected, but the Latvian goalie (and then AHL player) Kristers Gudlevskis stopped 55 shots and

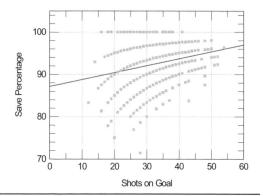

FIGURE 5.3. Save percentage of goalies during the 2013–14 NHL season as a function of number of shots received during a game. The line is a best linear fit.

allowed only 2 goals. He was brilliant the whole game. At the other end of the rink, Carey Price had cold feet: he received only 16 shots and was scored on once. And sure enough, it was on a breakaway. This was a bad combination for Price: a game where the stakes were high, the score was low, and there were few shots on goal.

It can be proven statistically that the more action a goalie receives, the easier it is for him to stay in the ideal mental zone and make saves. Take a look at figure 5.3, which compares the save percentage of NHL goalies in situations where they receive various numbers of shots. In the 1,000 NHL games compiled, each goalie played at least 55 minutes of the 60-minute regulation time. The linear fit shows a trend toward increased efficiency when the number of shots is high.[1] This trend has been verified to be real and not the result of some bias such as pulling a goalie early when too many goals have been scored. Instead, the data are reproduced well by a simulation model that assumes that the effectiveness of a goalie improves with the number of shots. In games with 40 shots or more, the average saves percentage is around 94 percent whereas it drops below 90 percent when there are 20 shots or less. The same pattern is observed whether the goalie is on the winning or the losing side.

1. Note that data points tend to align along curves. This is because both the number of shots and number of goals scored are integers, which implies that save percentages can only take certain discrete values.

These data dispel the idea that pummeling a goalie wears him down and makes him less effective. Of course, it doesn't mean players should limit their shots on goal, because the differences in save percentages don't make up for the chance each shot gives you to score. As the saying goes, a shot on goal is never a bad play.

Note in figure 5.3 the "one shot" save percentage of around 87 percent. This number is only what the trend suggests, an extrapolation, because games with only one shot on goal practically never happen. Still, it could be true that a not completely busy, "cold" NHL goalie would stop the shot about 87 percent of the time, a performance that would be considered inadequate by the standards of the NHL.

We noted earlier the similarities between the pitcher in baseball and the goaltender in hockey regarding their importance to the team. One key difference, however, is the lack of "specialists" in goaltending, unlike pitchers who come in categories such as starters, relief pitchers, and closers. Hockey goalies usually fall under two labels: "number one" for the one who plays most often and "number two" or "backup" for the other one. The difference is often just the overall reliability. To avoid overusing and tiring the number one goalie, the second goalie plays occasionally, sometimes against lower-ranked teams or when two games are played in two nights.

One may wonder why the better goalie is never brought in late into a game to finish it, like in baseball. After all, it could increase the chances of winning, say when a team is ahead by one goal. The answer is found in figure 5.3: a cold goalie, even a star goalie, coming on the ice in the middle of a game is probably not as effective as the warmed-up goalie who started the game. So, unless the sport of hockey develops a system akin to the bullpen where the non-playing goalie is kept warmed up and ready (which would defeat the purpose of giving him rest), in-game goalie rotation is unlikely to be used effectively.

When the sports media report on the previous night's hockey games, perfect games (shutouts) by goalies are always part of the highlights. Goalies who register a couple of them often get the title of Player of the Week or Player of the Month. Some years ago, the NHL felt there were not enough goals and too many shutouts in games. Since more goals make for better games to watch, the league took action to open up play. But it's a little-known fact that there are consistently many fewer shutouts in the NHL than there should be, statistically speaking. To give an example, the 2013–14 NHL season registered 138 shutouts, and given that the number of goals scored in a game follows a known statistical trend (as we will see in chapter 7), theory would have expected about 200 of them.[2] I can think of two reasons this might be the case. First, the team that hasn't scored puts more pressure on their opponents, especially near the end of the game, sometimes replacing their goalie with a sixth attacker. But I don't think this factor alone explains the low number of shutouts because even though a team's scoring chances are increased threefold when playing six-against-five, the majority of times the team still does not score in the 2-minute-long (or so) one-man advantage. Also, there is an added pressure to score only when the score is tight. When the score is lopsided, it makes for a rather dull endgame. The second reason—and the main one, I suspect—is mental pressure on the goalie. As I said, shutouts are the talk of the town the following day. So, in the mind of the goalie, the pressure is on him to preserve a perfect game, which increases as the clock is ticking. Nervousness sets in; the desire for perfection becomes a mental distraction. This is probably true at all levels of hockey. In my amateur career, I too have found the shutout to be rather elusive. I remember games where just when I thought a shutout was within reach, I'd be scored on in the dying seconds of the game.

Finally, let's examine another mental aspect of goaltending that is seldom talked about. One problem facing goalies is that their performance is scrutinized and compared game after game. Forwards and defensemen on a bad streak usually go unnoticed unless they go through an extended period of low productivity. If they score 2 points in one game and none the following two games, few analysts will criticize. By contrast, the number of goals scored against a goaltender is always prominently on display. This, combined with the fact that a goalie's performance is subject to large statistical fluctuations from game to game, can be difficult on the morale. As you can see in figure 7.2 for the case of Marc-André Fleury, a goalie can allow between zero and six goals in any given game just because of the inherent

2. The expected number of shutouts by a goalie is $S = Ge^{-GAA}$, where G is the number of games played and GAA is the goals against average.

randomness of the game. He could have a shutout followed by a six-goal game for no particular reason. It is a mental challenge to live through these ups and downs and not be affected by it. Experience certainly helps, and it is paramount for a young goalie to understand that it's not always his fault. Every goalie can attest that there are games where the puck seems to find every small opening to the net. In the next game, the goalie plays the same and looks brilliant, blocking most of the shots. Randomness in the game is the reason that it is difficult to correlate how prepared and physically well a goalie feels before a game and how many goals he will allow.

Are Bigger Goalies Changing the Game?

There's a trend of increasing goalie size in professional hockey leagues. NHL goalies taller than 6'3", like legendary Ken Dryden of the Montreal Canadiens, were once a rarity, but now we have Pekka Rinne (6'5"), Anders Lindback (6'6"), Ben Bishop (6'7"), Devan Dubnyk (6'6"), and a few others. Only 15 percent of NHL goalies are now less than 6 feet tall. The last drafted goaltender shorter than 5'10" was Lukas Mensator in 2002 (drafted 83th overall by the Vancouver Canucks). It looks like the days of professional goal-keepers the size of Arturs Irbe (5'8") and Darren Pang (5'5") are gone forever. In fact, Irbe was the last NHL goaltender shorter than 5'10" to play in the NHL (that was on April 4, 2004). Figure 5.4 compares the sizes of goalies of different eras drawn to scale and in relation to the net. Today's taller goalies block a significantly larger portion of the net.

Table 5.1 shows the average height of active goaltenders for three of the top professional hockey leagues. All the averages are above 6 feet. With a higher standard deviation, Russia's Kontinental Hockey League seems to accommodate a wider variety of heights, with the shortest goalie standing at 5'9".

Some experts blame the increasing size of goalies for the falling number of goals scored in the NHL. Figure 5.5 shows the number of NHL players reaching the typical 40 goal and 80 point benchmarks each season. Although there are some year-to-year statistical fluctuations, there is a general downward trend in the number of high scorers since the peak season of 2005–6, the time the NHL took actions to open up play and help scoring. While it is generally agreed

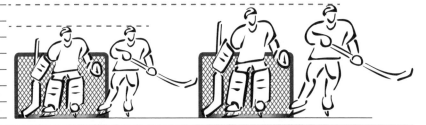

FIGURE 5.4. To-scale drawings of the shortest goalie of the 1980s and today's shortest player (both 5′5″ tall and about 5′8″ on skates) and today's tallest goalie and player (6′7″ and 6′9″, respectively, and about 6′10″ and 7′0″ on skates)

Table 5.1. Average height, with standard deviation, of goaltenders in major professional leagues, 2013–14 season

Swedish Hockey League (SHL)	Kontinental Hockey League (KHL)	National Hockey League (NHL)
73.5±1.5 inches	72.9±2.1 inches	74.1±1.9 inches

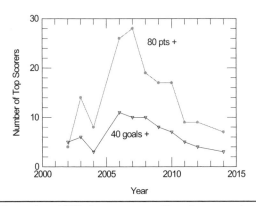

FIGURE 5.5. The number of NHL players to score 40 goals or more (triangles) and 80 points or more (full circles) during a regular NHL season

that players are fitter and better than ever, in the arms race against goalies, they seem to be on the losing side, at least for now.

As figure 5.6 shows, since 1998, the average goalie is 2 inches taller, and during the same period, he also gained 6.5 pounds. A similar trend is observed for newly drafted goaltenders, so it appears as though

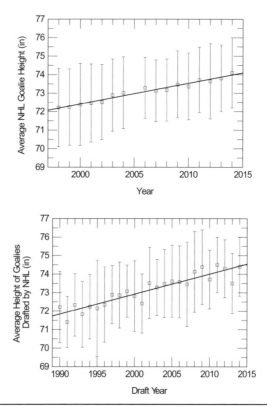

FIGURE 5.6. Average height of NHL goalies (*top*) and goalies drafted into the NHL (*bottom*). Bars represent standard deviation and the lines are linear best fits.

the phenomenon will continue for a while. The rate of increase is 1 inch per 9 years, which suggests that the average goalie will be 6′3″ tall by the year 2020, and several of them will be 6′7″ or taller. This is in re-markable contrast with the average size of defensemen and forwards, which has remained the same over the past 15 years (figures 5.7 and 5.8).

The increase in goalie size is partly caused by drafting more athletes from Northern Europe, where populations are generally taller. In a recent NHL roster (2012–13), North American–born goalies are, on average, 0.6 and 1.6 inches shorter than Finnish and Swedish goalies, respectively. This difference is close to that between the two populations. Still, North American–born goalies who were drafted have grown taller too: from 1997 until 2011, the average

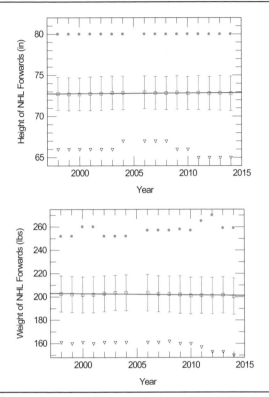

FIGURE 5.7. Height and weight of NHL forwards from 1998 to 2014 (with a minimum of 10 games played). Squares are the averages (with bars the standard deviation), and circles and triangles are the extremes. The line is a best linear fit.

Canadian- and U.S.-born NHL goalies grew by 1 inch. This rate of increase is much larger than the increase of height in the general population (about 0.4 inch per decade), so there appears to be a deliberate effort to recruit taller goalies.

Is this effort warranted? Are bigger goalies really better at making saves? Data comparing NHL goalie height, weight, and save percentages show no obvious correlation (figures 5.9 and 5.10). If anything, there appears to be a slight disadvantage in being taller or heavier, but the trend is too weak to make a definite call.

The conclusions we can draw are: (1) bigger NHL goalies don't fare any better than smaller ones and (2) smaller goaltenders can still compete at the highest levels. Every young goalie today who has

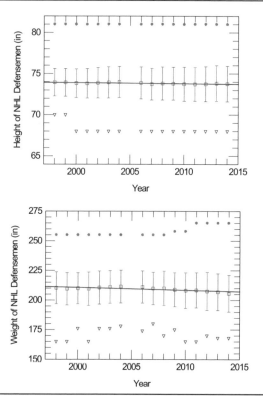

FIGURE 5.8. Height and weight of NHL defensemen from 1998 to 2014 (with a minimum of 10 games played). Squares are the average (with bars the standard deviation), and circles and triangles are the extremes. The line is a best linear fit.

aspirations to play in the big leagues should keep this in mind. After all, Tim Thomas (5′11″), a two-time Vezina winner for the league's best goaltender, was his team's most valuable player when the Boston Bruins won the 2011 Stanley Cup.

Having made that point clear, I will now say something that sounds contradictory: all else being equal, bigger goalies probably have an advantage over smaller ones. How can this be? We just saw that there is no correlation between size and save percentage. The reason is this: when we compare data among NHL goalies, we are not looking at a random sample of goalies but at an elite subgroup. All goalies who reach that level have comparable abilities regardless of whether they are smaller or taller.

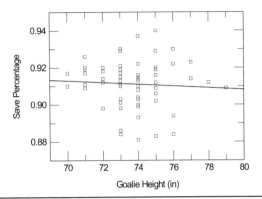

FIGURE 5.9. Relationship between height and save percentage among NHL goalies. The line is a linear fit, and data are from goalies who played at least 10 games in the 2011–12 NHL season.

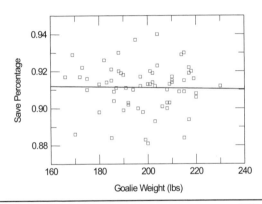

FIGURE 5.10. Relationship between weight and save percentage in NHL goalies. The line is a linear fit, and data are from goalies who played at least 10 games in the 2011–12 NHL season.

But if we were to look at all goalies of all levels, the bigger ones would have an advantage because of two reasons: (1) shots often come too fast to react and (2) shooters are not perfectly accurate. Yes, snipers look good when they score amazing goals, but we should not forget that they fail to score 90 percent of the time. Many shots miss the net or hit the goalie directly, an easy save. The limited shooting accuracy is most apparent at the All-Star Accuracy Shooting competition, where the best shooters will often need several attempts

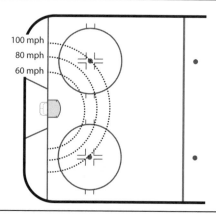

FIGURE 5.11. Shots taken within a certain range cannot be stopped by human reflexes (assumed here to be 0.2 second). The three half-circles are the minimum stoppable distances for various puck speeds.

before they hit a circular target 1 foot in diameter placed some 20 feet away. The direct implication for goalies is that covering more area of the net is an advantage. Also, while quick reflexes certainly help, sometimes they are not enough. There is a limit to human reaction times, around two-tenths of a second, as we have seen, so when a quick wrist shot is taken from the slot, there would not be enough time to react. One can imagine a fictitious "circle of doom" around the net within which the goalie's reaction times are insufficient. The concept is depicted in figure 5.11. For an average shot of 80 mph (130 km/h), the radius of this circle is 23 feet (7 meters) away. Puck deflection is a common occurrence and a goalie's reflexes are usually not fast enough. When a powerful shot is taken from the blue line by a defenseman, the goalie generally has enough time to position himself to make the save. But the shooter's teammates will often try to deflect the shot with their sticks—something that requires quick hands but that is routinely done with amazing skill—making it impossible for the goalie to adjust in time.

On a close-range shot, the best a goalie can do is take a posture that covers the net as much as possible. And since the legs are about twice as slow as the arms, they better get moving ahead of time. This is the idea behind the popular butterfly technique: the goalie goes down and puts his pads parallel to the ice, covering the bottom portion of the net, the most vulnerable area. Figure 5.1a shows an example of the so-called blocking butterfly variant used for close shots. The glove, blocker, and body are quickly put into the "pre-set" position shown. When the shot is taken from far away (figure 5.1b), the butterfly technique is still used (again because the legs are slower and more vulnerable) but the hands are mobile to block any puck shot in their direction. This variant is sometimes called the "reaction butterfly."

The success of so-called positional goalies, like Carey Price, relies on a precisely mastered technique of "looking big" from the shooter's point of view and moving around the net while keeping the same effectiveness from all shooting angles and situations. Because of its effectiveness, this style of goaltending is now the norm, and the bigger you are (while being still athletic), the better it is.

In the end, however, no matter what the numbers say, if a small goalie can stop 90 percent of the pucks, the scouts are likely to take notice. NHL teams are multimillion-dollar businesses after all, and winning is how they generate revenue. Given this reality, it is unlikely that the trend of favoring tall goalies is the result of some bias by hockey scouts and team managers, but it cannot be completely ruled out.

Does Having a Star Goalie on the Team Make Much Difference?

As you will see in the final chapter of the book, the effectiveness of NHL goalies is remarkably uniform across the league. The save percentage in the league as a whole averages 0.910, and the bulk of goalies' scores are between 0.900 and 0.920, a difference of only plus or minus 1 percent. How much difference does this make in a game?

In a typical game, 30 shots are fired to the net by each team. So a difference of ±0.01 in save percentage translates into a difference of only ±0.3 goal, not even a full goal. In other words, a star goalie with a save percentage of 0.920 would make 1 extra save every 3 games compared with the average goalie (1 extra save every 90 shots, on average). Since many games are won by more than one goal, is this slim advantage to be discounted?

As amazing as it sounds, the difference between +0.3 goal per game and −0.3 goal per game is roughly 10 games won or lost at the end of an 82-game season. To give an idea of what this means, it's typically what separates the team standing in first overall position in the league and the team standing at the middle of the pack. So an above-average goaltending performance can make the difference in terms of the team's participating in the playoffs or not.

To explain why this is the case, let's start with the simple fact that the number of goals scored on each side in a typical game averages around 2.7. Given that the number of goals scored follows a Poisson distribution—a way of examining how often a given event occurs in a given time frame (see chapter 7)—we can calculate that the win percentage of a team drops by 15 percent for each increase of 0.3 goal scored against them. So a difference of 0.6 goal per game changes the win percentage by 30 percent. This is a statistical trend, not a rule, so sometimes a goalie with an above average save percentage will have a lower win percentage, and vice versa. There are aspects of the game over which the goalie has no control and that are a reflection of the team's defensive play as a whole.

When a Goalie Makes the Difference

The ultimate feel-good experience for a goalie is to win by a score of 1-0, because this is when he knows for sure that he made the difference. Were it not for the shutout, the team would have tied at best, or lost.

A goalie's performance is commonly measured by numbers like the save percentage, the number of wins, and the goal against average (abbreviated GAA, the average number of goals scored against a goalie per game since the beginning of the season). These are fairly reliable measures of performance over the long run. But can we measure the impact a goalie has had in any given game?

To illustrate what I mean by impact, let's consider a game where the goalie does poorly but the team wins, say by a score of 8 to 7. We could say he was saved by his team and his poor performance had no real impact in the end. But if the team couldn't step up to compensate and win, then we can say that the goalie made a difference, negatively. Let's take the opposite situation: a team plays poorly and scores only two goals but the goalie plays well and only lets one goal in. Then he saved the day and made a positive difference. On the other hand, if his team played very well too and won, say, 8-1, then the goalie's good performance didn't matter much in the end. In other words, a goalie's fluctuation in performance doesn't always have an impact, good or bad.

We can identify games that were influenced by the goalie's performance, positively or negatively, in the following way. Based on the number of shots received in a game, we know how many goals should be scored, on average: it is the number of shots multiplied by the goalie's save percentage over the season (as we have seen, this is typically around 0.900 in the higher hockey leagues). Of course, this expected number of goals is not always equal to the number of goals actually scored.

Let's take an example: Carey Price receives 30 shots in a game, he gets scored on 4 times, and the Canadiens win 6-4. How did Price's performance affect the game? (Of course, we're answering the question from a purely statistical point of view, from game data only, and not from watching the game. In reality, a goalie can make several key saves and be a determinant factor, regardless of his save percentage during the game.) With an average save percentage of 0.900, we would have expected 3 goals in 30 shots from a typical NHL goalie, so we can say that Price gave up 1 extra goal. Since the team won by 2 goals anyway, however, his subpar performance was nullified. If he had let in only 2 goals (1 less than expected) and the team had won 3-2, then his performance would have had a determining positive effect: by letting in one goal fewer than usual, he made his team win.

In this kind of analysis, there are different possible scenarios, summarized as follows:

1. Let's call A the number of goals above normal scored on a goalie (rounded to the nearest integer for simplicity). If the team loses by N goals, the goalie's performance had a determinant negative impact if $A \geq N$.
2. Let's call B the number of goals below normal scored on a goalie (rounded to the nearest integer for simplicity). If the team wins by N goals, the goalie's performance had a determinant positive impact if $B \geq N$.
3. In all other scenarios, the goalie's impact is nil.

How many times does each scenario occur in reality? Table 5.2 analyzes a sample of 5 games played by Carey Price during the 2011–12

Table 5.2. Analysis of impact of goaltender Carey Price (0.913 save percentage) in five games of the Montreal Canadiens during the 2011–12 NHL season

Opposing team	Outcome	Goal differential	Shots against	Game save percentage	Goals against	Number of goals above (+) or below (–) average	Impact on game
Toronto	5-0 win	5	32	1.000	0	−3	neutral
New Jersey	2-1 win	1	29	0.966	1	−2	positive
Florida	1-2 loss	−1	34	0.941	2	−1	neutral
Toronto	4-5 loss	−1	27	0.815	5	+3	negative
Winnipeg	0-4 loss	−4	34	0.882	4	+1	neutral

Note: The number of goals above or below average is determined from the average save percentage of all NHL goalies during that season (0.914).

NHL season. Over the entire 2011–12 season, Price kept the net 65 times for the Montreal Canadiens. On 10 occasions, his subpar performance had a negative impact, but on 14 occasions his better-than-expected performance had a positive impact. For the other 41 games, his performance did not impact the outcome of the game.

How Hard Is Professional Goaltending?

It's been suggested, half-jokingly perhaps, that to enhance the experience of watching the Olympics, they should include ordinary people among the athletes, just to have a point of comparison. We could say the same about hockey. How would an ordinary hockey player fare in an NHL game? Would he ever score or collect assists? Conversely, how would a professional player fare in an amateur league? If anything, it would be a great practical joke to have a star player like Sidney Crosby play a game in beer league, just like NBA star Kyrie Irving, dressed up as "Uncle Drew," participated in a local pick-up basketball game. He showed how a "70-year-old" can dribble through a crowd and slam dunk the ball. As I am writing these lines, I hear that Chicago Black Hawks' Patrick Kane (dressed up as himself) reaped 10 points in an amateur hockey game in Buffalo. In the same line of thought, have you ever wondered what would happen if you put an amateur goalie in net for a professional hockey game? How good, or how bad rather, would an average goalie do compared with a pro?

Such an experiment has never been done at the top levels of hockey, but it's happened in a league not too far below, during an official game of the Ontario Hockey League (OHL), one of Canada's top junior leagues. Connor Crisp, who normally played center for the Erie Otters of the OHL, got the unexpected request to keep the net when the team's two goalies were injured. With Crisp as goaltender, the team lost 13-4 to the Niagara Ice Dogs. To be fair, we should say that he also made 32 saves, so his 0.711 save percentage is not bad for a complete rookie who, visibly, was not quite at ease with his newly borrowed goalie equipment. But it's also possible that the Ice Dogs eased off on him once the win was secured.

The best strategy for a complete beginner put in such a situation would be to stay in front of the cage and hope the puck bounces off you. The area of the net covered by a fully dressed adult male goalie is about 60 percent. Oftentimes the shooter's best position to score from is straight in front of the net because the net looks biggest from that point of view. At angles between 50 and 60 degrees from that optimal line of view, the net is tilted, looks smaller, and is about the same size as the goalie. This would be the case when shooting from the faceoff dot, roughly.

As a beer league goalie, I also had a chance to face NHL shooters. This happened not in a game, of course, but during the shooting of a documentary called *Scoring with Science* with the Toronto Maple Leafs in 2012. Acting as a co-host to Jay Ingram (former host for the Discovery Channel's *Daily Planet* show), I was asked to describe, among other things, how it feels to face a hard slap shot from the big league. In the leagues I play in, the slap shot is not even allowed, so it was a big step for me. So I anxiously but happily faced the shots by NHL defenseman Jake Gardiner and forward Matt Frattin (seen in figure 5.12), and I confronted a few of their breakaway and deke attempts as well. Shots were taken from the slot, some 20 feet away, at speeds up to 85 mph, as measured by radar located behind the net.

My sole objective was not to look too bad and try to make at least one save, which I was able to make more often than I had

FIGURE 5.12. NHL defenseman Jake Gardiner (*left*) and forward Matt Frattin (*right*) with the author (*center*) during a filming for the Discovery Channel in 2012

anticipated. At the highest speeds, however, my reflexes were not good enough: by the time I moved my glove, the puck was already in the net.

Maple Leafs' goalie Jonas Gustafson had faced the shooters before me and had fared much better, needless to say. He confirmed to me that on close shots sometimes there is not enough time to react. All you can do is move your body a little here and there, but sometimes that's enough to make the save. Some will say that tall goalies like Gustafson, who is 6'3" (3 inches taller than I am), cover a lot of net, which helps. This is true, but a solid technique is also absolutely necessary.

When facing NHL-caliber shots like I did, with the player standing still and shooting at a precise moment, the difficulty level is relatively low but not representative of real games. When the game is unfolding at blistering speeds, with players cutting across the net, making quick passes and unexpected shots, this is where the professional goaltenders shine.

From limited data and my own experience, my estimates on how an amateur goalie like me would do in the NHL are as follows. An amateur would probably attain a save percentage in the neighborhood of 0.500 to 0.700. That's not bad, you may say, but wait a minute. With an average of 30 shots per game, that translates into between

10 and 15 goals, and sometimes more. The team would simply have no chances of winning.

Do Video Replays Help the Game?

The rules of hockey say a goal is scored when a puck completely crosses the red line on the ice that connects the two goal posts. In other words, when viewed from above, some "white" from the ice should be visible between the puck and the red line. If the puck is inside the goalie's glove, the whole glove must be seen to cross the line for the goal to count.

It sounds simple enough. But although referees and the goal judge (the one turning the red light on when a goal is scored) often make amazingly accurate calls on goals, sometimes things happen too quickly for the eye to see, and this is where video replays are important. Video replays are sent to judges (located in Toronto, for the NHL) who review the play from different angles.

As it turns out, most video replays are not done to check whether or not the puck crossed the line but rather to ensure that no rule violation took place before and during the goal. Here's a list of common reasons for a goal to be voided:

- The puck went in after the time expired on the clock (precise to the tenth of a second).
- The net was knocked off its pegs before the puck went in (unless it was judged to be done intentionally by a member of the defending team, in which case the goal still counts).
- The puck was kicked or pushed into the net deliberately by a member of the attacking team.
- The puck was touched by an attacker's stick at a height above the cross bar prior to entering the goal.
- The puck touched or deflected off a referee or any ineligible player before entering the net.

There are also other reasons to deny a goal, such as goalie interference and pushing the goalie into the net after a save, but these are subjective calls made by the referee, and they don't call for video replays.

Video replays are made from different camera angles and in slow motion. Particularly useful is the camera positioned high above the net in the rafters. From this vantage point, the vertical cross bar and the red line are usually seen parallel and side to side. Other cameras monitor from the sides and the back of the net. There is also a camera inside the net, usually in the top middle portion of it (the "Netcam"), and sometimes in the boards behind the net. By remote control, the net camera is rotated to look around. Because it is exposed to direct impact, it is built to withstand pucks, but it is not unbreakable. It has a special type of lens that can capture the entire net opening at once. The only drawback to such a wide field of view is that it makes the posts appear slightly rounded, an effect called "barrel distortion," commonly associated with wide-angle lenses.

In addition to making the game fairer, camera replays capture events that are sometimes unusual or just plain puzzling. Take for instance the astonishing "miracle save" by Jonathan Quick, a feat that seemed to defy the laws of physics. In a 2011 game between the Calgary Flames and the Los Angeles Kings, a hard shot was fired from the side by Jarome Iginla. The Kings' goalie appeared to have gotten a piece of it but not enough to stop it from going past him and toward the empty net. From the top camera view (a drawn screenshot is shown in figure 5.13), the puck seems to hit Quick's blocker, then proceeds to travel straight toward the open net. But, lo and behold, before it gets there, the puck bounces off thin air and changes direction, missing the net. Was there an invisible wall? Was God a fan of the Los Angeles Kings? After all, they did win the Stanley Cup the following year. So what happened? (Before you read on, you may want to look at the replay on the Web and try to solve the puzzle for yourself.)

From a physics point of view, a mass bouncing off nothing in midair is not possible. So the puck must have hit something heavy. A possibility (not verifiable from the top camera's view) is that the puck went down and hit the ice. Looking at the side-view camera, this is what appears to have been the case. The change in direction, as viewed from above, coincided with the puck hitting the ice, at which point it started traveling parallel to the red line.

But this does not explain why the puck changed its path when it hit the ice. Because ice is slippery, there is little horizontal force provided by the ice to make the puck change direction horizontally.

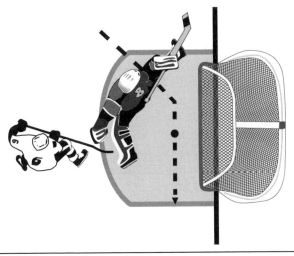

FIGURE 5.13. Top view of the "miracle save" by Jonathan Quick. The dotted line shows the trajectory of the puck that was partially blocked by Quick. (Illustration by Ovide Theriault)

Without such horizontal force, the puck should have kept going in a straight line as viewed from above, bouncing off the ice and entering the net. But enough force was imparted to the puck by the ice to kick it away from the net.

Ice has been known to play some tricks on the puck. In 2006, Pittsburgh Penguins' goalie Sébastien Caron missed a puck, shot from far away and bouncing on the ice and that somehow veered to his left and into the net. Ice friction (see chapter 1), acting on a puck that is bouncing and flipping, will sometimes give it momentum in some random direction. In the case of Quick's miracle save, the puck ricocheted off his blocker (or stick) and shot downward at great speed and slammed against the ice. This could have made the friction force on the ice all the more important, enough to deflect the puck away from the net. It is also likely that the puck acquired a fast spin after hitting Quick's blocker, which would have helped increase the impact force against the ice. It's all theory, of course, but whatever happened that night was, thankfully, caught on camera, and while it was very unlikely, it makes for an awesome physics problem.

When to Pull the Goalie

There's a rule of thumb hockey coaches use that says that when the team is trailing by one goal, the goalie should be pulled and replaced with an extra attacker near the 1-minute mark of the end of the game. When trailing by two goals, the goalie should be pulled around the 1:30 or 2:00 mark. This rule applies to 5-on-5 situations or while playing with a one-man advantage. When the trailing team is playing shorthanded, the goalie is rarely pulled.

The motivation for replacing the goalie with a sixth attacker is that there is a greater chance to score a goal and tie the game. In the NHL, a team produces a goal once every 27 minutes, on average, when playing 5-against-5. When playing 6-against-5, the team with the extra player will score once every 8.5 minutes[3]—a 3 times greater chance to score than usual. But it is understood that the chances of getting scored against is increased without a goalie—the rate is 1 goal every 3 minutes or so. But it's still worth it: if the team is going to lose, losing by one more goal makes no difference, while the increased chance of tying the game is a bonus. Another benefit of putting an extra attacker in is that it tends to draw more penalties from the defending team that is under pressure, and when it happens, the prospects of playing 6-against-4 are even better for the trailing team.

The statistics have been analyzed for many different scenarios in a paper by David Beaudoin and Tim Swartz. As it turns out, given the frequency at which goals are scored when there are different numbers of players on the ice, the current strategy of pulling the goalie at the 1-minute mark is not optimal. In a situation where a team is trailing by one goal and playing 5-on-5 (the most common scenario), pulling the goalie at the 3-minute mark until the score is tied would be a better option. When the team is trailing by 2 goals, pulling the goalie at the 6-minute mark is better than pulling him at the 1:30 mark. These two examples show that the coaches don't decide optimally.

There are probably reasons that coaches don't pull goalies at earlier times than what is generally accepted. Pulling the goalie at

3. See D. Beaudoin and T. B. Swartz, "Strategies for Pulling the Goalie in Hockey," *The American Statistician* 64 (2010): 197–204.

the 6-minute mark and getting scored on could draw criticisms from the fans and the media, even if it were the best strategy. Of course, if it did work, the coach would look good, but more often than not the team would lose. The benefits of it working once in a while does not outweigh the problems of coming under criticism most of the time.

We have now looked at the rink, the ice, the skates, the sticks, and the players. When you combine all these elements into a game of high speed and intensity, there is always a chance of injury. We will explore common injuries and what might be done to prevent them in the following chapter.

The Injury

It's been suggested that the only sport in which players absorb heavier blows than hockey is the good old medieval game of jousting. It's indeed the only contact sport in which players travel faster than they do on skates—thanks to the horses, of course.

Full-body contact hockey can be a rough game indeed. Hockey players move faster than running athletes in other team sports, including soccer and American football. And the blow from a body check depends more on speed than weight: double the weight and the impact force doubles, but double the speed and the impact quadruples. Because of their high speed, hockey players carry more kinetic energy (the energy carried by moving masses, which is equal to the energy necessary to bring them at their speed from rest) than athletes from any other contact sport. Two mid-sized hockey players skating at moderate speed who collide on ice will dissipate enough energy to power a 60 watt lightbulb for a minute and a half. So, yes, it is possible to severely injure another player in a high-speed collision. There have been more than a few accounts of death and near-death incidents following violent collisions on the ice. Luckily most of the injuries are not that serious.

Protective Gear

Hockey players wear a complete set of equipment for good reasons. Even in amateur leagues, where body checking is not permitted, there is still plenty of danger from flying sticks and pucks and from knee and elbow hits. Figure 6.1, which shows two players slamming into the board, demonstrates the importance of wearing protective gear against skate blades and hockey sticks.

Unlike mounted knights, hockey players don't wear metal armor. They need light but flexible protection against cuts and blows. The goalie's mask, for example, is a thin (3 millimeters thick) shell of

FIGURE 6.1. Dallas Stars' Brenden Dillon checks Chicago Blackhawks' Andrew Shaw into the board. (AP Photo/Charles Rex Arbogast)

fiberglass or carbon fiber and sometimes Kevlar. It is constructed to resist the full impact of pucks. The inner padding is made of soft polymer-based foam for comfort and to absorb blows. The back of the mask is a moveable cup fastened with stretchable straps, so that the mask has some mobility. The design is very effective, and goalies almost never get injured when hit on the mask by a puck, even though this does happen quite a lot.

Items as simple as hockey jerseys have gone through rounds of innovation and improvements too. The ideal hockey jersey does not retain heat (it "breathes") and repels water. It also needs to be flexible so as not to impede on a player's movement. In an effort to improve performance and comfort, the NHL has adopted a standardized jersey made of a material that is supposedly more aerodynamic and retains less water.

Types of Injuries

In spite of improvements in equipment and protective attire, injuries are frequent occurrences in hockey games. This is partly because the rules allow for body-checking a player that is carrying the puck. An average hockey player skating at top speeds carries more kinetic energy than bigger athletes in sports like American football. At any given time in a season, an NHL team can be missing half a dozen players due to injuries.

When it comes to injuries, we all have a sense about what dangers to avoid. Although we may not be able to describe them in terms that are technical and scientific, the things we watch for are problems related to pressure, strain, and acceleration.

Too Much Pressure

Pressure injuries are easiest to understand. They happen when force is applied over a small area of the body. That's why the bullets from a gun are dangerous: not because they carry a lot of energy—in fact, one experiences little recoil when shooting or receiving a bullet—but because they concentrate a great deal of force over a small area on the body, more than it can withstand. In hockey, examples of pressure injuries are bruises and bone fractures caused by a hockey puck that impacts a small area of the body. (We will look at some numbers on this below.) Another example is the sharp edge of a skate blade, which cuts by applying pressure on the skin. Other common types of pressure injuries in hockey include spearing and slashing by hockey sticks.

To protect against pressure injuries, hockey equipment uses hard shells to diffuse the blow. Elbow pads, chin pads, and helmets are good examples. They have hard external shells and soft inner padding. If a player is struck in the head by a puck or a stick, or falls to the ice, the impact force is diffused by the helmet to a larger area of the body. Helmets are so effective against such hits that the player often feels no pain at all. As a goalie, I have received hard shots to the helmet and experienced little more than a loud bang and sometimes noticed an odd smell of burned rubber from scuffmarks by the puck.

To prevent cuts from skate blades, layers of soft fabric can work just as well as hard shells. For example, thin, cloth-like neck guards are worn by many young players, and, although they don't protect against pucks and sticks, they do prevent cuts. Cuts to blood vessels and to the neck area are particularly dangerous. Injuries to muscle tissues and tendons are not life-threatening, but they can be potentially disastrous to a hockey player's career. The tissues that make up tendons and ligaments (the ones connecting bones at joints) are under strain several times higher than the body's weight during normal body movements. As a result, damaged tendons take a long time to heal and may never recover their full strength. Following high-profile incidents of skate blade cuts, including those involving NHL defensemen Andrei Markov and Erik Karlsson (severed Achilles tendon) and Joe Corvo (lower leg cut), cut-resistant socks and wrist guards made of cloth with Kevlar have gained in popularity among players.

Feeling the Strain

Strain and compression injuries are quite common in hockey, and they are unfortunately the hardest to prevent. Many times, they are self-inflicted. Strain injuries occur when body segments move out of their normal range, that is, either too fast, too much, or the wrong way. During a mid-ice collision with another player, for example, a player's legs may swing and become hyperextended, causing strain in the knees or the groin area. Goalies extending their legs—doing the splits—sometimes feel the sting of a pulled groin. The strains may only cause temporary discomfort, but at other times they result in damaged tendons and ligaments, and the injury may last for months afterward. Strained or torn anterior cruciate ligaments (ACLs) is a common problem in hockey and other sports.

Injuries also occur when body parts become compressed together. When a hockey player is checked from behind (which is forbidden by the rules of the sport), the backward motion of the head causes the cervical vertebrae to squeeze against each other, potentially bruising the padding in between, the intervertebral disks. Helmets can't prevent such injuries. The resulting compression is the same as the "whiplash effect" in a car accident. Even during a legal body check, parts of the player's upper body, like the ribcage and collarbones, may undergo considerable stress.

If hockey equipment were to be designed to protect against strain injuries, they would have to limit the range of motion of players' legs, knees, arms, and neck. This would conflict with the need of hockey players to move freely and easily, however. Of all types of injuries, strain injuries are probably the ones where the improvement is most needed, but it would require radical innovations. Attempts have been made in the protection of knees, for example. Braces are sometimes worn to strengthen knee joints that have been weakened or damaged. They are also used as precautionary measures on healthy knees, although their effectiveness in that regard has been debated.[1]

Acceleration Injuries

Acceleration is a change of velocity divided by time. A well-known example is a car accelerating from zero to 60 mph as the driver goes from a stop to highway speeds. The less time it takes, the larger the acceleration. In hockey, all is well when speed changes smoothly, like when a player races for the puck or brakes to a stop. In such cases, the body deals with accelerations that are on the order of half a g. (The g stands for the gravitational pull of the Earth and its effect on how fast something can accelerate; the standard measure is 1 $g = 9.8$ m/s².) But when velocity changes abruptly, like during a collision between players or when hitting the board or the ice during a fall, acceleration is at its highest, and that is a problem.[2] Internal organs and the brain can experience damage when they are quickly accelerated (or decelerated, which has the same effect). Most collisions in sports like hockey are largely inelastic, meaning that players don't tend to bounce off each other but lose their speed. This implies that the kinetic energy is absorbed during the shock by the players and their equipment.

During a heavy blow to the head, accelerometers placed in the helmet have measured as much as 20 to 30 g of acceleration.[3] Another study found accelerations ranging from 10 to 100 g[4] and lasting for about 5 milliseconds. (In comparison, the typical accelerations we experience every day in cars, airplanes, and elevators are a fraction of

1. B. G. Pietrosimone, T. L. Grindstaff, S. W. Linens, E. Uczekaj, and J. Hertel, "A Systematic Review of Prophylactic Braces in the Prevention of Knee Ligament Injuries in Collegiate Football Players," *Journal of Athletic Training* 43, no. 4 (2008): 409–15.

2. This is also understood with the concept of impulse: to produce the same change of velocity in a shorter time period (a shorter collision time), a greater acceleration is needed and a greater force must be applied.

3. K. M. Guskiewicz and J. P. Mihalik, "Biomechanics of Sport Concussion: Quest for the Elusive Injury Threshold," *Exercise and Sport Sciences Review* 39 (2011): 4–11.

4. R. S. Naunheim, J. Standeven, C. Richter, and L. M. Lewis, "Comparison of Impact Data in Hockey, Football, and Soccer," *Journal of Trauma: Injury, Infection, and Critical Care* 48 (2000): 938–41.

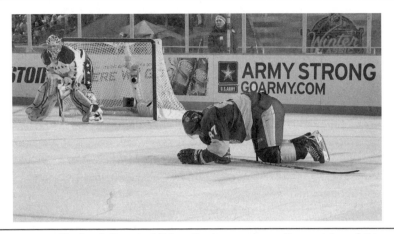

FIGURE 6.2. Pittsburgh Penguins' Sidney Crosby goes down after receiving a hit to the head during a game against the Washington Capitals. The concussions suffered by star hockey players have helped attract attention to this recurring problem in hockey.

one *g*.) These are the accelerations of the outer parts of the skull, also called "surface acceleration." Inside the head, the self-cushioning of the brain reduces the acceleration. High accelerations are the reason that concussions happen, a problem that has plagued not just hockey but also other contact sports like football, soccer, and boxing. Some incidents involving star players have generated debates on how to prevent concussions. An example from hockey was when the Pittsburgh Penguins' Sidney Crosby (see figure 6.2) had back-to-back head hits and was sidelined for several months. There is also a great deal of media attention on concussions in football and the lasting and sometimes fatal effects on its players.

To prevent concussions and other acceleration injuries, there is no other known solution than effective cushioning. The laws of physics establish an inverse relationship between acceleration and cushioning length: increase the padding thickness, and the acceleration drops accordingly. This is the idea behind the crumple zone at the front and the back of a car. Car frames are designed to squish during a frontal impact while the passenger cabin remains rigid. Car safety engineers say that the crumple zone absorbs the shock of impact, but in reality what it does is allow the passengers to stop over a longer distance. Longer distances mean longer stopping times and less acceleration.

In crash test videos, dummies are seen moving by a meter or more even though the front of the car has hit a fixed wall. The longer the stopping distance (the longer the crumple zone), the weaker the acceleration of the passenger will be. The same idea motivated the introduction of soft acrylic windows and boards specially designed to move on impact in NHL rinks. The added displacement may appear small, but players say that it feels softer than slamming against a sheet of tempered glass affixed on a board that won't budge. It may also prevent injuries in situations like figure 1.4A, where player Nathan Gerbe (at 5′5″ and 178 pounds) is hit by a heavier Milan Lucic (at 6′3″ and 235 pounds).

Similarly, the padding inside the hockey helmet is meant to soften the blow by acting like a crumple zone. But the problem is this: the typical helmet padding is poorly designed for that purpose. It may feel smooth and comfy to the touch, but if you try to press into the inner padding, it is nearly impossible to do so. In some helmets, layers of hard polystyrene foam make it downright impossible to sink your thumb into it. Now imagine how much force it would take to squeeze this padding with your head. To squeeze the whole centimeter or so of padding, you'd have to fall on the ice from very high, in which case the helmet would be useless anyway. So why is the foam padding there in the first place? For comfort, period.

When a player's head collides with something hard like the ice, it stops instantly because of the rigid inner padding. The acceleration's effect on the brain is likely to result in a concussion. To be sure, the skull won't crack or suffer any external bruises because of the very effective pressure diffusion by the external hard shell, but if an accelerometer were to be located inside the head, it could not tell whether the head has hit the ice with or without a helmet. The two situations have the same effect on the brain, and the brain has only itself as a cushion.

Is it possible to improve hockey helmets? Probably. Professional football leagues have made progress using air pockets and softer materials to reduce concussions. Even baseball helmets have better designs than hockey helmets, with thick and very spongy layers of foam. In hockey, Bauer introduced special liners specifically designed to minimize impacts and rotational accelerations, which were found to be especially damaging to the brain. But even in the best possible designs, there is not a lot of room—only a few centimeters at most—to play with cushioning. There are many practical and aesthetic reasons that helmet manufacturers can't put thick padding in hockey helmets. But going from the current situation to helmets with a few centimeters of softer, more efficient padding would make a difference. It could be done by choosing foam materials with hardness such that it would squeeze completely only during an impact of magnitude high enough to cause a concussion.

Just How Dangerous Is a Hockey Puck?

Many sports involve kicking, throwing, or hitting a projectile, whether it is a golf ball, baseball, football, or badminton birdie. But projectiles in sports are not equal in their capacity to injure. One important element, as discussed in the previous section, is the area of impact: small, hard objects tend to concentrate pressure and to cause bone to fracture. A powerful kick may send a soccer ball at good speeds, but due to its size and softness the ball usually delivers little more than a painful slap (although it has been shown to cause concussions too in some cases). A baseball, in contrast, is more dangerous because it is smaller and harder and travels at speeds of 100 mph or more. Even more dangerous are golf balls: good golfers drive them at 150 mph at tee off. Thankfully, when people do get hit by them, they are usually far away on the field, so the ball has lost much of its speed because of air drag.

Hockey pucks stand near the top of the list in their capacity to inflict injuries. The disk of hard, frozen, unforgiving rubber is only 1 inch thick. Experiments have found the amount of pressure required to cause cranial fracture to be about 15 times the atmospheric pressure.[5] A hockey puck can deliver that much pressure to a head at velocities as low as 30 mph. To give an idea of that speed, it would take a puck 4 seconds to travel from one end of the rink to the other. This could be a typical passing speed, attainable by 12-year-old players and adult goalies. Keep in mind that the 30 mph figure is an approximation: from shot to shot, impact pressure may vary by as much as

5. H. Böhm, C. Schwiewagner, and V. Senner, "Simulation of Puck Flight to Determine Spectator Safety for Various Ice Hockey Board Heights," *Sports Engineering* 10 (2007): 75–86.

25 percent. The exact impact location on the head is important too, as the skull is not evenly thick, the temples being more vulnerable than other areas, for example. Of course, we are not saying that cranial fractures by pucks are common in hockey, thanks to the helmet, but it's to give an indication of the damage a puck can do.

Hitting a player is not the only cause for concern with flying pucks. Think of what could happen if a projectile traveling 100 mph struck a fan in the stands. This threat has been examined thoroughly and resulted in various safety measures, such as glass panels put around the rink to protect spectators. End zones also include nets put atop the glass to stop pucks from reaching fans sitting behind the goals. Tall windows surrounding the rink provide a twofold benefit: (1) to stray into the crowd, the puck needs to travel high, and, in doing so, it loses speed to gravity and air drag; and (2) pucks shot at high angles (above 20 degrees) tend to be slower, a consequence of the mechanics of shooting. It is easier to shoot fast near the ice than in the air.

Thanks to effective equipment and rink designs, it is remarkable that mishaps with a puck account for only a small fraction of all hockey injuries, to both athletes and spectators.

Throwing Off the Gloves

On a winter night in a small Canadian town, the local hockey arena is filled with a loud and cheering crowd, fueled on French fries and beer. The home team is playing its archrival, a team from a town just half-an-hour's drive away. They've played each other many times before and they know each other well. An ambulance waits near the arena's entrance. This is the kind of hockey league where fans expect to see as much physical play as skills—and a few fights. Players target the puck carrier, not the puck, and they hit to hurt.

The home team is losing 4-1, and, with only a few seconds left on the clock, the home team enforcer—a tough volunteer, mostly—goes to work, to give the crowd something to cheer for. He drops the gloves and, with the three referees looking on passively, he exchanges punches with an opponent. Less than a minute later, the opponent loses his footing, and the local player falls on top of him, becoming an instant hero, the crowd chanting his name. At least the game ended on a positive note, they would later say. The crowd disperses peacefully, players exchange jokes in the hallway. They can't hang around too long because they have day jobs too.

To an outsider who doesn't know about hockey, a game description like this would seem strange. It is odd indeed, for what other sport in the world gives only a 5-minute "major" penalty for engaging in a fistfight, something that could under normal circumstances land you in jail?

Some say fighting in hockey exists out of necessity. Roughness is unavoidable, and much of it is actually unintentional. The physicality makes emotions flare. Even skilled players like Mario Lemieux, Wayne Gretzky, and Sidney Crosby had fights at some point in their careers. Without fighting, one theory goes, any mediocre player could take a star opponent out of a game with a mid-ice collision, an elbow to the head, or a knee-on-knee hit. Having to answer to the team's enforcer acts as a deterrent.

Others argue that fighting in hockey is not a necessity but has all the characteristics of a cultural phenomenon, something that is learned from watching other players, something that is passed on from one generation of players to the next. In many leagues, including the NHL, fighting is not allowed on paper but tolerated within a certain set of unwritten rules, a kind of "fighting etiquette," so to speak. There is an honor code. For example, punching a player who is injured or who has fallen is considered off limits.

With so many fans enjoying watching the fights and roughness in hockey, there is an economic incentive for leagues to allow it. (Studies have found that the more enthusiastic a fan is about hockey, the more likely he is to support fighting. The people most opposed to fighting tend to be those who don't watch hockey, although there are plenty of hard-core hockey fans who oppose such violence.) The NHL would probably lose revenue if it were to eliminate fighting, and other than the additional 2-minute penalty for instigating a fight, the league has done nothing to reduce fighting. On the contrary, roughness is an asset for commercial exploitation. At the glitzy American Airline Center in Dallas, I saw a pregame montage on the big LED screen that could not have been outdone by Hollywood: video snippets of fights, scuffles, and bone-crushing hits—all real-game examples by Dallas Stars' players—accompanied with thumping music. It was quite effective in getting the crowd worked up. In a region of the country where hockey is not so well established at the grassroots level, this kind of display certainly helps attract new fans. But on the other hand, perhaps it leaves the wrong impression that hockey is just about that, hitting and fighting, and the occasional goal.

Young hockey players sometimes receive contradictory messages with respect to roughness. At the 2011 World Junior Championship, a member of Canada's National Junior Team gave a solid body check from the side to a Czech player who was coming through center ice. The hit made him spin half a turn before falling unconscious onto the ice. The crowd erupted with loud cheers. Later, when the injured player was immobilized and carried away on a stretcher, a somewhat more somber crowd politely clapped and whistled, as if to say: "no one wants to see somebody getting hurt." Minutes later, when the referees handed a penalty to Team Canada, effectively giving 5 minutes of power play to the Czech team, the crowd booed loudly. In a context where it was not certain whether the injured athlete would ever play again, it was a disrespectful thing to do. And to a young and impressionable mind, this was a confusing message, to say the least.

In many hits, there is no real intent to injure, but in some instances, precautions could have been taken by the perpetrator. Sport psychologists know that mental preconditioning is an important element of what happens on the ice. A split second is all the time a player has to decide whether to deliver a clean body check, or aim for the head, or lift his elbow, and so on. An argument often heard in defense of a hitter is this: "I didn't intend to hit his head and injure him. He just had to keep his head up." While this may be true, the state of mind of the player delivering the hit at that precise moment has a determinant effect on his decision, and this is influenced by his learning of the game, the instructions he's received, in the dressing room and from his peers, what he has learned is okay or not, condoned or not.

Deliberate blows to the head are serious concerns to professional hockey leagues and to other sports. According to the rules, players should make every attempt to avoid hitting a player who is not carrying the puck. Earlier I talked about Sidney Crosby's experiences with repeated concussions. One of them took place at the Winter Classic of 2011, where David Steckel seems to have made no visible effort to avert a collision with Crosby, who received a blind side hit to the head from Steckel's shoulder. This was probably an avoidable incident, and, for the league's top player, it started Crosby on a long series of concussion problems and many months of recovery. In recent years, the NHL has been stricter in applying suspensions for blows to the head and body checks from behind, with some tangible results. All questionable hits are now reviewed by a committee, and suspensions are given when no visible efforts have been made to avoid collision with the other player's head.

The negative effects of repeated blows to the head have been known for a long time by the medical community. Repetitive fighting, like some enforcers do for a living, can damage the brain. As far back as 1988, the Canadian Academy of Sport Medicine took a clear stand on fighting and body checks. They proposed that fighting be eliminated completely and body checking not be allowed in those minor leagues that are not meant to be training schools for professional leagues.

Hockey has been around for more than a century, and this aspect of the game may not change in the near future. At the end of the day, the sport will be what the people want it to be. But we know that it doesn't have to include cheap and dangerous hits. Some of the best hockey ever played, with the best combination of skills and determination to win, is at the Winter Olympics, like during the 2010 gold medal match between Team Canada and Team USA and the 2014 gold medal game between Canada and Sweden. The games were physical, but skill and intensity, not roughness and violence, were the driving forces.

If some of the roughness and dangers of hockey could be avoided to some extent, there is another element of the sport that is unavoidable and has an even greater influence on the outcome of games: randomness. In the following chapter we examine randomness and its effect on the chances of a team winning or losing, and how it influences the production of players and goalies.

THIRD PERIOD

The Final Score: Who Wins,
Who Loses, and by How Much?

CHAPTER SEVEN

The Odds

Having looked at many technical aspects of hockey, from the ice to the equipment, and how the best players master the sport, we now turn to what really matters—at least to some: winning and losing. For that, we need a few tools from statistics, for how can you play the numbers without using them?

Picking a Winner

Predicting which team will win a hockey game is never a sure thing, but that does not prevent people from doing it every day. Some, like sport analysts, bookies, and gamblers, do it by profession, but most of us do it just for fun. We are the hockey fans, hockey pool players, and buyers of sport lottery tickets. And when it comes to making predictions, there is no shortage of advice. Some look at numbers and statistics, sometimes lots of them; others rely on factors that can never be measured precisely, like team spirit and momentum.

But what does science have to say about our ability to predict the outcome of a hockey game? In answering this question, we need to consider two facets of the problem. The first has to do with the way we can use statistics to estimate the odds of one team winning over another. The second aspect is less obvious, perhaps deeper, but just as important: given the statistics and hockey knowledge that is available out there, what is the best success rate we can hope to achieve in picking a winner?

More often than not, our expectations about which team should win are shaped by the overall league standings. The number of games won, or the team winning percentage, is the simplest and most popular indicator of team strength. The team with the highest win percentage is generally considered the favorite. We can estimate the odds of one team winning over another based on the assumption that, if they play in the same league against the same teams, their success rates against the league as a whole are indicative of strength, just like we would compare the strength of two persons by having them lift the same barbell. So let's suppose Team A and Team B have win percentages w_A and w_B, respectively. Given the above assumptions, the following equation estimates the odds of Team A winning the game:

$$p = \frac{w_A(1 - w_B)}{w_A(1 - w_B) + w_B(1 - w_A)} \tag{1}$$

This equation is similar to one given in my previous book, but this one works in a wider range of situations. The formula essentially compares the odds of Team A winning any given game against the odds of Team B losing any given game. (A more detailed justification is given in appendix C.) Also, win percentages—shown in the equation as w_A and w_B—are expressed in decimals (with values between 0 and 1), that is, by writing 0.700 instead of 70 percent, for example.

Using equation 1, we find that a team with a 0.600 record should win 60 percent of the time against a team with a 0.500 record and 70 percent of the time against a team with a 0.400 record. When teams have equal records, the odds are 50-50, so you might as well toss a coin to pick a side. Note that equation 1 behaves correctly when one of the teams has a 1.000 or a 0.000 record but not when they both have 0.000 or 1.000 records.

There is a word of caution to be said about using win percentages early in a season, however. When few games have been played, a single game may cause big swings in the win percentage, and this is an indication that it is not yet a reliable measurement of team strength. In statistics, small samples mean big problems. (I will state this over and over; it is extremely important when doing statistics in any field.) One way to circumvent this problem is to use the records from the previous season as a starting point, and then gradually incorporate the new season's statistics.

In addition to winning percentages, other known factors influence the chance of a team's winning a game. For example, there is home ice advantage. Teams tend to win more at home than on the road. It's not a strict rule, as some teams actually win more on the road, but in a league like the NHL, teams typically win between 0 and 60 percent more games at home than on the road. During the 2011–12 season,

56 percent of all games in the NHL were won by the home team. This translates into win percentages at home that are 1.12 times higher than average and win percentages on the road that are 0.88 times smaller than average. Since all teams play the same number of games on the road over the whole season, the advantage cancels out during the regular season, but in a playoff series, starting at home is a definite edge.

But home ice advantage has a positive effect in hockey leagues because it means that the majority of fans attending games get to see their favorite team win. Happy fans mean good business—a win for everyone.

We take into account home ice advantage by simply using the home and away win percentages in equation 1. But again, beware of the small number of games problem, even more so here because we're splitting the stats between home and away games. In some cases it might be best to use the overall win percentage of a team and fudge it by a factor that reflects the usual home-away difference for the league as a whole. To illustrate, suppose Team A plays at 0.450 overall and hosts Team B playing at 0.550 overall. Using the figures above, we may adjust w_A to $0.450 \times 1.12 = 0.504$ and w_B to $0.550 \times 0.88 = 0.484$, thereby giving Team A a slightly better than 50 percent chance of winning.

Another factor that helps or hurts a team's chances of winning is the two-games-in-two-days scenario. Teams don't like to play two nights in a row, and with good reason, as the following table demonstrates. The 2005–6 regular NHL season had 448 games where the two-in-a-row format occurred (table 7.1), and the consistent tendency is a below-normal performance by teams playing on the second night. The only time a team plays better in the second game is when it returns home from a road trip. Overall, in all scenarios combined, the second night sees the win percentage dropping to a factor of about 0.8.

Yet another factor of influence is the number of games in a row a team has won or lost. In my previous book, I reported the surprising fact that winning and losing streaks tend to be shorter than expected in the NHL. By "expected" I mean compared to the typical streak length obtained in a situation where successive games don't influence each other. Since the finding, I looked at this effect from different

Table 7.1. Win percentages by teams that played in two games
on two consecutive days

| Game 1/ Game 2 | Win percentage | | Change in performance in Game 2 (%) |
	Game 1	Game 2	
Home/home	75.0	56.3	−18.7
Home/away	67.2	38.1	−29.1
Away/away	46.9	41.9	−5.0
Away/home	36.2	48.3	+12.1

angles and crunched more data to make sure it is not some artifact of misused statistics. For example, we might ask whether the reason is the difference between wins in home and away games. Is it possible that winning streaks starting at home are cut short when the team goes on the road, for example? This hypothesis was tested by incorporating it in a simulation and it showed no measurable effect. Or could it be what we commonly call the "law of averages," or the idea that streaks have to be broken so that the team's winning percentage remains close to where it should be? Alas, there is no such law in statistics; it is just a widespread misconception.

Instead, all additional data confirmed the original observation: winning and losing streaks do have a tendency to end prematurely. Said differently, and perhaps more shockingly: after it wins, a team tends to perform worse than it usually does, and after a few losses, it tends to perform better than it usually does. Figure 7.1 shows the win percentage of NHL teams after having played a winning (or losing) streak of some length. The win percentage is expressed in terms relative to its normal performance, so that 1 represents a team's usual (normal) performance. The dashed curve is the result of a random simulation of 82 games where the outcomes of games don't influence each other. In this simulation, the computer generates series of 0s and 1s with a probability matching the winning probability of a team. The streaks are then sorted by length and the average length is calculated.

There is a remarkable difference between the simulation and real-life data. After 3 losses, a team wins more often by a factor of 1.16; so if it plays normally at 0.600, it will play at 0.700 after losing 3.

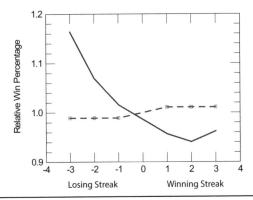

FIGURE 7.1. Effect of streak length on a team winning percentage. Negatives are losing streaks. The dashed curve is the expected result when successive games don't influence each other.

Moreover, the effect is asymmetric: it is stronger on the losing side. Whatever impetus makes a team play better or worse does not act with the same intensity when the team has lost or won.

The effect of streaks on a team's winning chances is somewhat counterintuitive. When a team loses a few games in a row, one would think that it would be demoralizing and that the spiraling down effect would make losing even more likely. That's why, according to popular belief, you have to "shake your team up" in a period of drought. A trade, a firing, or a good scolding may do the trick. Conversely, if a team is on a roll, you'd think that boosted morale would make the team even stronger. But this viewpoint doesn't stand up to scrutiny.

Let's now look into the matter of prediction accuracy. Given all the hockey statistics and hockey knowledge that is out there, is it possible to pick a winner with close to 100 percent success rate? If not, what's the best one could hope to achieve?

I'll start with an anecdote. In 2007, a friend and I developed a model to predict the winner of NHL games. We did this as an experiment, for fun, and yes, I'll admit, with the faint hope of making money too. The model was purely based on game data and did not involve human judgment. The model took into account past performance, relative team strength, home ice advantage, and the two-games-in-a-row factor. This is all standard stuff, of course, but the novelty was this: instead of using the win percentage as a measure of

strength, teams were continuously rated in the same way chess players are—by the Elo rating system. Each team was assigned a rating
that was adjusted after each game. But with Elo, a win is not just a
win: the reward (the amount of rating adjustment) is higher when a
team beats a strong team than when it beats a weaker one; the same
idea applies to losing. In addition to its simplicity, the Elo rating
system has the advantage of taking into account the strength of the
teams played against, so it tends to eliminate the artificially inflated
performance of teams playing in weaker divisions, for example. In
terms of prediction accuracy, our expectations were conservative: we
thought a 70 to 80 percent success rate might be reasonable. Boy, were
we in for a surprise!

Feeding the model with daily game results and making rating
adjustments accordingly, the model successfully predicted the winner in only 56 percent of games over the entire 2007–8 NHL season.
At the start of the season, when teams were not well rated, the rate of
success was lower, but it improved over time. Of course, some games
were easier to predict than others: in games where team strength dif-
fered the most, the success rates hovered around 70 percent, but in
most games, our success rate barely beat that of a coin toss. Sometimes a stronger team would lose at home against a weaker team that
had played the night before—every bit of data our model used favored
the stronger team—so there was no way the model could have made
the correct prediction. In hockey, like other sports, there is always
an element of uncertainty.

Needless to say, the model did not win us any money with sports
lottery because to break even we needed a 72 percent success rate. Like
playing at the casino, the payouts are just too low so the odds are in
favor of the house. The statisticians who calculate the payouts for lottery corporations know how to be on the safe side. (That's not to say
money can't be made when betting against other people. In gambling,
every bit of knowledge helps. But gambling was not of interest to us.)

We learned three things with this experiment: (1) it's not easy to
make accurate predictions in a league like the NHL; (2) there's no
sure way to make money with sports lottery; and (3) we tend to overestimate the future performance of a model, especially when the
model is optimized with the results from past games. It's known
among scientists that sometimes a really good fit to data is obtained

when several parameters of a model are adjusted at the same time. Or, to put it in non-scientific terms, hindsight is always 20/20. Part of the problem has to do with the fact that the model adjusts to the "noise" of past games, or the unavoidable random element of the game. This "hindsight" effect leads to artificially high anticipated success rates.

So why is it difficult to accurately predict a winner? We now recognize that the main difficulty is the parity between professional hockey teams. Teams are more evenly matched than many fans would admit. For example, during the 2011–12 season, the winning percentage of the American Hockey League (AHL) teams ranged from 38 to 72 percent in the East and from 41 to 59 percent in the West. The majority of teams were within ±7 percent of playing 0.500 (that is, their win percentages ranged from 43 to 57 percent). In the same year, the NHL Eastern Conference teams ranged from 38 to 62 percent and the NHL Western Conference teams from 35 to 62, with most teams within ±8 percent of the average. The teams of the 2007–8 NHL season (the one we modeled) were particularly evenly matched, to within ±6.6 percent of average. With such a narrow range of winning percentages, games are tightly disputed and randomness plays an important role. In fact, by using equation 1 to calculate the odds of a team winning, and assuming a league where teams are distributed to within ±6.6 percent of 0.500, the theoretical success rate in predicting a winner would be 57 percent, close to the result we obtained with our model. An approximate rule says that the success rate of a model cannot be much higher than 50 percent plus the league's spread (standard deviation) on winning percentages.

In professional hockey leagues, most teams have a fair chance of winning any given game, and this is even more the case during the playoffs, where only the top teams get to play against each other. Yet, when hearing comments from fans and analysts, one would think that the outcome of a playoff series is as certain as the sun rising tomorrow. With that mindset, it's no wonder that we call the sixth-ranked team eliminating the top team an "upset." But, looking at the statistics, the chances of this happening were not that small after all.

But there's a positive side to all of this. Team parity in a league is generally a good thing, and sport leagues recognize that. The NHL has drafting rules that help teams at the bottom catch up to others.

In a league where teams would vary a lot in strength, predictions would be easier to make but many fans would lose interest. The element of uncertainty makes the sport interesting to watch.

Skills versus Chance

If we were to rank games according to how much chance plays a role, on the one hand we would have chess and checkers, games of skills and strategy, and at the other extreme, we would have games of pure luck, like rock-paper-scissors and snakes and ladders (but don't tell that to your toddler when he wins). Sports, including hockey, would fall between the two extremes. No sport is only about skills or only about chance.

That's why a team that is considered to be weaker can beat a stronger one. Statistics and models can't predict the random events happening in a real-life situation. We don't know where a given contest will go in the next few seconds, let alone an entire game. Sometimes a puck zips through the legs of players and finds the back of the net; other times it is deflected away by someone's skate. And how many times have we seen a goalie in a technically correct position only to see the puck squeeze through a tiny opening? That's why a game result might go against the team that the statistics favored. I like the example of a game played on November 5, 2011, when the slumping Boston Bruins, then standing last in their conference, visited the Toronto Maple Leafs, then standing first overall. Few would have predicted that night that the Leafs were to lose 7-0 on their own ice.

When people use a word to characterize what science is about, they often use "accuracy." Science is mathematical and accurate, in contrast with many areas of life that are fuzzy. So when we use hockey statistics in a systematic, scientific way, we tend to focus on finding small and subtle advantages of one team over another or one player over another. Yet, scientists are all well aware of the limitations of their tools when dealing with the uncertainties of real-life situations. As a matter of fact, one of the most useful numbers that emerges from sports data is the amount by which we cannot predict the outcome of games and the amount by which we cannot differentiate teams and players. Because of randomness in hockey, there is an unavoidable amount of fluctuation in the performance of teams and players from game to game, the so-called noise we talked about earlier. Numbers like points per game, goals per game, and other types of productivity measurements continually sway on either side of the values we would consider "normal" or "expected." Sometimes it seems that the numbers are so far off from normal that we wonder if there's something unusual going on. Or is it just luck—good or bad? Science can help us answer this question. The laws of random events can help us decide how much fluctuation is to be expected.

In the short run, random fluctuations can make any team or a player look very good or very bad. In the long run, however, the difference in skills will manifest itself and the best will rise to the top of the standings. As we saw above, this effect is known as the law of small numbers: over a small number of games, large deviations from normalcy should be expected. In statistics, small samples mean big trouble as far as predictive reliability goes.

This is not something purely theoretical. We all have a gut feeling for the law of small numbers. After a few games into a new hockey season, no one is surprised to see a lesser-known player at the top of the scoring list. But on other occasions, we get fooled by the same phenomenon. For example, it's not rare to see sports writers turn a team's short losing streak into a headline story.

So how much fluctuation is to be expected in the game of hockey? There are tools to help us decide when a series of events have an underlying cause and when they are random, tools used by scientists to study the randomness found in Nature.

One such statistical tool is the Poisson distribution, which predicts how many times events will occur over a given a period of time. (By the way, the name of this tool comes from its discoverer, Siméon Poisson [1781–1840]. Poisson means "fish" in French; it is somewhat ironic in that it is also used to study fish population distributions.) The Poisson distribution can answer questions like "How many shutouts and how many 5-goal games should we expect from a goalie with a 3.00 goals against average, or GAA?" These questions would be hard to answer with gut feelings only.

Figure 7.2 shows an example of the application of the Poisson distribution. The performance of Pittsburgh Penguins' goalie Marc-André Fleury over the 2010–11 NHL season is broken down by the number of goals he allowed in a game. The results are compared with

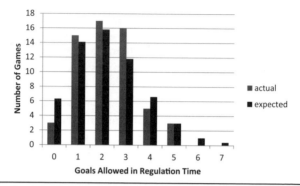

FIGURE 7.2. Number of regulation time goals allowed by Pittsburgh Penguins' goalie Marc-André Fleury during the 2010–11 season (2.24 GAA)

the statistically expected number of games for each number of goals allowed, calculated from his respectable 2.24 GAA. Only the goals allowed in regulation time and only the games he played in their entirety were counted, for a total of 59 games and 132 goals. Of course, the predictions are not identical to the actual numbers, but they do follow the same trend. Although allowing five goals to slip by him would be considered a performance on the weaker side, it appears to be normal to have this happen a few times in a season. Also notice how Fleury had only half the number of shutouts expected from theory, in accordance with what I discussed in chapter 5.

A second example of the Poisson distribution is given in figure 7.3, this time for the point productions by forwards. Here the performances of Vancouver's Daniel Sedin and San Jose's Logan Couture are broken down by the number of points (goals and assists) they scored during a game. Sedin was the NHL's top scorer in 2010–11 with 104 points in total and 1.27 points per game (PPG), and Couture was one of the top rookies with 56 points and 0.71 PPG. Again, theory and actual numbers follow a similar path, with the notable exception that Sedin could perhaps have had a few 4- and 5-point games. It is possible that he may have had a tendency to let his teammates take the lead after he scored a few points.

Let's now apply the same statistical tool to answer a more general question: what should the distribution be of goals scored against a

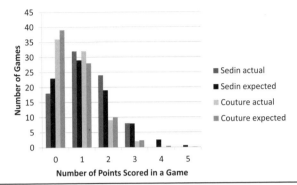

FIGURE 7.3. Number of points scored by Daniel Sedin of the Vancouver Canucks (82 games played) and Logan Couture of the San Jose Sharks (79 games played) during the 2010–11 season

Table 7.2. Expected number of games by number of goals allowed for goalies with various goals against average

Number of goals allowed	Goals against average				
	1.50	2.00	2.50	3.00	3.50
0 (shutout)	13.4	8.12	4.92	2.99	1.81
1	20.1	16.2	12.3	8.96	6.34
2	15.1	16.2	15.4	13.4	11.1
3	7.53	10.8	12.8	13.4	12.9
4	2.82	5.41	8.01	10.1	11.3
5	0.85	2.17	4.01	6.05	7.93
6	0.21	0.72	1.67	3.02	4.63
7	0.05	0.21	0.60	1.30	2.31

Note: The number of games played is 60.

goalie who plays 60 games, the typical seasonal workload by a starting goalie? Table 7.2 gives the number of games by the number of goals expected for goalies with different GAA. Notice how the GAA has a particularly strong effect on the number of high-scoring games. Readers may also be surprised to see as many 5-goal games as shutouts for a goalie with a respectable 2.50 GAA. If this is correct, should we have to question a goalie each time he allows that many goals? Probably not.

Table 7.3. Expected year-to-year fluctuation in point production by a team or by a player

Typical point production over a season	Expected fluctuation	
	Points	Percentage
110	±12	±11
100	±12	±12
90	±11	±12
80	±10	±13
70	±10	±14
60	±9	±15
50	±8	±16
40	±7	±18
30	±6	±20
20	±5	±25
10	±3	±30

Goalies are not the only ones to experience random fluctuations in performance. Players and teams do, too, even when they consistently try their best game after game. In table 7.3, we list the typical fluctuation in point production by a player over an 82-game season. The results are sorted according to the expected point production, which is something that we can normally estimate from production in previous years. Here, "point production" may be any measurement we wish to make: passes in a season, points in a season, goals in a season, and so on.

If we compare players with different productivity, we find greater fluctuation in points at the top. A player who scores 100 points one year may score 112 or 88 points the following year for no particular reason. Percentage-wise, however, the trend is the exact reverse: it's ±30 percent for a production of 10 points per year and ±10 percent for 100 points per year. Randomness has a greater effect on players who score only occasionally. Again the law of small numbers is at play.

Next, we tackle a question of great importance to hockey teams, managers, coaches, scouts, and even hockey pool players: how long should we wait before we can decide with confidence about the strength of a certain player or a team? This is a question we face each time we have a new player on a team or a new team in a league. In a

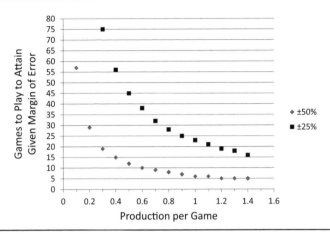

FIGURE 7.4. Number of games to be played to attain a margin of error of ±50 percent and ±25 percent on the team's production per game

situation with no data available, we have to remember that random fluctuations after just a few games can make an average player look like a star . . . and a star look like an average player.

Figure 7.4 is a guide to help us decide when we can know whether a person or a team is a flash in the pan. The horizontal axis gives the production per game observed up to a given point. The vertical axis gives the number of games to be played before we can be sure that the observed production is correct to within ±50 percent or ±25 percent. For example, a ±50 percent margin of error on a production of 0.5 PPG is ±0.25 PPG, so we would expect the production to stabilize between 0.25 and 0.75 PPG in the long run.

To show how the results of figure 7.4 can be used in many different problems, let's tackle a few practical questions inspired from real-life situations.

1. A rookie player scores 6 points in his first 12 games. How confident are we that this 0.5 PPG production will be kept over the entire 82-game season? Answer: According to figure 7.4, an uncertainty margin of ±50 percent is attained after 12 games for PPG=0.5. In other words, we should expect his long-run PPG to be between 0.25 and 0.75. So after 82 games, we should not be surprised to find his production ranging between 21 points and

62 points (or 41 points plus or minus 50 percent). In other words, the uncertainty is still relatively high.

2. A player recently called up from a farm team has scored 4 points in 3 games. How significant is this production? Answer: For PPG = 1.3 we'd need at least 5 games to attain a modest ±50 percent confidence, and 18 games to be sure within ±25 percent. The 3 games played are insufficient to be sure within ±50 percent.

3. During the 2010–11 NHL season, Toronto Maple Leafs' forward Phil Kessel had a point production of 0.8 PPG. The following season, he led the league with 30 points after 22 games (1.4 PPG). Some analysts said it was a fluke. How likely was this caused by just randomness? Answer: At PPG = 1.4 the confidence attains ±25 percent after 16 games. After 22 games range of uncertainty is even smaller. So it is reasonable to consider that his increased production is the result of some real improvements, personal and/ or team-wise, and not just chance. (As it turned out, Kessel went on to maintain this higher performance for the following four seasons.)

We can do a similar analysis for the point performance of teams. There is a key difference between team points and player points, however: team points come by units of one (for ties or overtime losses) and two (for a win), whereas players may score between zero and several points in a game. As a result, team point production resembles that of flipping a coin (or a loaded coin rather), and this calls for a different statistical tool called the "binomial distribution."

Figure 7.5 has been created using the binomial distribution. The horizontal axis gives a certain production per game, such as the win percentage (or the tie percentage, or any such quantity of interest). The vertical axis shows the number of games required before a margin of error of ±0.200 and ±0.100 is attained on that production.

Let's apply the results to a realistic scenario. A hockey team traditionally known to be playing around 0.500 has started the season with a rather mediocre 2-7-3 record (wins-losses-overtime losses). The win percentage is 0.167. Is there something to worry about? Answer: According to figure 7.5, with a production of 0.200 (rounding 0.167), we need at least 22 games to attain a ±0.100 margin, so the 12 games played are not enough. To attain a ±0.200 margin of confidence, we

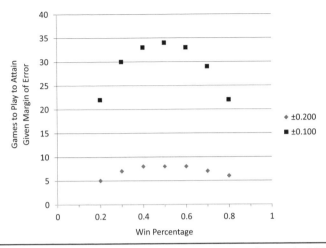

FIGURE 7.5. Number of games to be played to attain a margin of error of ±0.200 and ±0.100 on a team's production

need only 5 games. Estimating the actual margin after 12 games to be roughly ±0.150, we could expect the long-run win percentage to be not much higher than 0.300. So it is reasonable to expect the team to underperform in comparison to previous years.

To conclude this section, a disclaimer is in order. The numbers given in this section can't substitute for observation, hockey knowledge, and judgment. They are meant to give estimates for the normal range of fluctuations that occur in a sport like hockey. In doing so, assumptions are made that no other key factor is at play. For example, it is possible for a player to experience a production drop not because of randomness but because of injury, personal problems, or problems with teammates and coaches. A team as a whole can also suffer for concrete reasons. But when nothing unusual has happened, the numbers quoted in this chapter are indicative of the statistical fluctuations we should expect.

Of course, many more types of analysis could be done, and different confidence levels calculated for the myriad of specific situations that happen every day in hockey. But to do so, we would have to study the statistical tools in depth, which is beyond the scope of this book. (For those of you who want to know more about the math and statistics of hockey play, my earlier book contains further information.)

When to Panic (and When Not to): The Surprising Statistics of Bad Streaks

When people worry about their hockey team, it is often because the team is not winning, a player is underperforming, or the goalies let in too many goals. Sometimes there are reasons behind below-normal performances. Other times, there seem to be none, and this is when it is useful to know how much bad luck is to be expected and at what point we may want to try to find an underlying cause. In this section we consider the statistics of sequences of poor performances by players, teams, and goalies. Results are broken down according to the expected (or "normal") production by a player, a team, or a goalie. But since the expected production itself is a guess, an estimate normally based on past productions, the following statistics are meant to be interpreted as guides only.

Player Production

Every hockey player goes through periods of meager production at some point in a season. Sometimes there is a specific reason for it, but not always. So here's a question of practical importance to hockey team managers and coaches: how much production drop is normal, and at what point do we say that something must be wrong?

We will consider two scenarios. The first scenario is a "hard slump," that is, when a player does not score at all (and has no assists) for a number of games in a row. The second scenario is a "soft slump," where production does not shut down completely but hovers below expectation. Both cases are treated with the same statistical tool used in the previous section.

Table 7.4 gives the longest streak without production that should normally happen over a period of two complete NHL seasons, or 164 games. This period of time is selected so that the quoted streak length would have a 50 percent chance of occurring in a single season. Scoreless streak durations are given for players with different expected production. According to the results, a star player may produce 100 points per season (1.2 points per game), yet a 4- or 5-game slump would not be extraordinary. For a typical 40-point producer (0.5 PPG) we may wait 9 games or more before ringing an alarm. It could also be con-

Table 7.4. Longest sequence of games without production (goals, assists, or points) expected to happen with a 50 percent probability in an 82-game season

Normal production (points per game)	Longest slump duration (no. of games)	Normal production (points per game)	Longest slump duration (no. of games)
0.1	25	0.8	6
0.2	16	0.9	5
0.3	13	1.0	5
0.4	10	1.1	5
0.5	9	1.2	4
0.6	7	1.3	4
0.7	7	1.4	4

Table 7.5. Average number of streaks without production to be expected over an 82-game season

Hard slump duration (no. of games)	Per-game production (points per game)		
	0.5	1.0	1.2
1	8	12	12
2	5	4	4
3	3	2	1
4	2	1	0
5	1	0	0
6	1	0	0
7	0	0	0

sidered normal if a 30-goal scorer (0.37 goal per game) played up to 10 games in a row without finding the net.

Another relevant question is the number of slumps we should expect during a regular season. Table 7.5 shows the number of streaks without production for players who have a given production per game. To take one example, we can roughly expect 1 sequence of 6-game drought from a player producing 40 points per year (0.5 PPG).

This is all theory, of course, but how do these numbers compare with real-life statistics? In figure 7.6, three representative examples are compared with statistical predictions. The three players were ran-

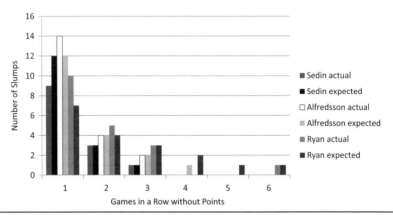

FIGURE 7.6. Comparison between predicted and actual number of slumps by players over an entire NHL season

domly chosen to represent different levels of production, and in each case the predictions of theory are close to the results.

Let's now look at soft slumps. According to the laws of statistics, productivity (measured in production per game) should fluctuate, and the amount of fluctuation depends on the period of time (number of games) we are considering. In other words, we can't simply consider productivity fluctuation all by itself; we need to consider the number of games over which it takes place. For example, when a player normally produces 1 point per game, on average, for him to score a total of 2 points over 10 games is not the same as scoring 4 points over 20 games, even though both situations give the same 0.1 point per game. The second case is more worrisome, as we will see.

Figure 7.7 shows the amount of fluctuation in the production (as percentage of the expected production), predicted by statistical theory. The numbers can be used to estimate the lower limits for poor performances and the upper limits for high performances.

I should stress that the fluctuation percentages given in figure 7.7 are not hard boundaries. In other words, we cannot say that all performances falling outside the limits must be a problem. Instead, they are meant to give us an idea of the typical amount of uncertainty. For example, when we expect ±20 percent fluctuation in performance and observe a 23 percent drop, it doesn't necessarily mean there is an

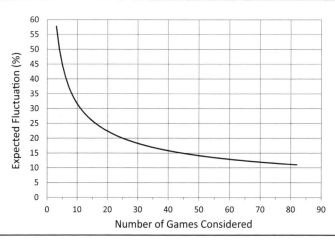

FIGURE 7.7. Amount of productivity fluctuation predicted by statistical theory

underlying cause. But the farther the productivity is outside the suggested range, the less likely it is caused by chance alone.

With the results of figure 7.7 we can now answer practical questions like the following:

1. Player X normally produces 25 points in an 82-game season, or 0.30 PPG. Near the end of the NHL season, as the team is fighting for a place in the playoffs, Player X has scored only 1 point in his last 5 games. Analysts are pointing fingers at him. Is it justified? Answer: Over a 5-game stretch, productivity could be expected to deviate by ±45 percent according to figure 7.7, or ±0.14 PPG for Player X. His productivity is expected to vary from 0.16 to 0.44 PPG, and his recent 1 in 5, or 0.2 PPG, falls within that range. There should be little cause for concern, at least not yet.

2. Over the past 5 years, Player Y has scored an average of 80 points per season (40 games per season in his league). Midway into a new season, he has only 20 points. Can he claim bad luck? Answer: Player Y normally produces 2 PPG, and this productivity may vary by ±22 percent over 20 games. His current productivity has dropped by 50 percent, more than twice the expected amount. Chances are there are reasons other than bad luck behind this drop in production.

3. Player Z is a third-line center who typically scores 30 points in an 82-game season (0.37 PPG). He scores 6 points in the first 8 games of the season. The coach considers promoting him to the first line. Would it be justified? Answer: Over 8 games and from an expected 0.37 PPG production, we would normally see a ±35 percent fluctuation, or ±0.13 PPG. His current 0.75 PPG production is much higher than the upper limit of 0.40 PPG, and thus it is likely that there is real improvement, not just chance.

Team Production

We now analyze the problem of team winless streaks, a problem that tends to spill even more ink in the press than when individual players don't perform. In hockey-crazed cities, media and fans are quick to call for action. After just a couple of bad games, talk of trading players or firing coaches can be heard. (Of course, once things turn around, the team becomes a Stanley Cup contender.) The question we should ask is this: how long should we wait before shaking things up? Although this can be a difficult question, and should never be answered with numbers alone, the laws of statistics can help the decision-making process.

Let's first look at hard slumps by hockey teams. Figure 7.8 shows the longest winless streak that is expected in a season by a team. The criterion used here is the same as we used for players: the quoted longest streak would have a 50 percent chance of happening over an 82-game season, or once every 2 seasons on average. Only win percentages between 0.300 and 0.700 are shown because, in competitive hockey leagues, teams tend to be that close in strength. (For example, at the end of the 2010–11 NHL season, 13 out of 30 teams had winning percentages between 0.450 and 0.500, and 24 were between 0.400 and 0.600.) As we see from figure 7.8, a team that typically plays 0.500 could lose 6 games in a row, and this would not be out of the ordinary.

Next we look at soft slumps. The question here is this: given a number of games played, how low can the win percentage go before we say there is a problem? Three factors need to be taken into account here: (1) the win percentage we expect from the team, (2) the number of games under consideration, and (3) the confidence level

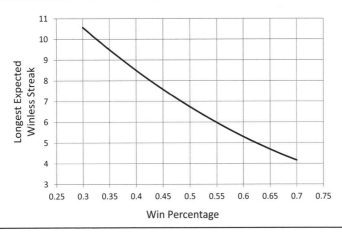

FIGURE 7.8. Longest winless streak predicted by statistical theory to occur in a season with a probability of 50 percent

we want. In the following statistics, I will use confidence levels of 90 and 70 percent, and I will explain what they mean later.

In table 7.6, we find the minimum number of games we can expect a team to win over a given number of games played. Results are sorted by (expected) win percentages and confidence levels. To give an example, a team that normally plays at 0.500 should win at least 6 games out of 15 consecutive games with a 70 percent confidence level, and at least 5 games with a 90 percent confidence level. This means for any sequence of 15 games played in a row, the team should win 5 games or more 90 percent of the time, and 6 games or more 70 percent of the time. The table can also be used to find the odds for winning even fewer games. For example, if a team has a 70 percent chance of garnering at least 5 wins, it has 100 − 70 = 30 percent chance of winning 4 games or fewer.

Here are some practical questions we can now answer:

1. The Boston Bruins finished the 2010–11 season with a 0.561 win percentage and went on to win the Stanley Cup. At the beginning of the following season, they reaped only 4 wins in the month of October (out of 11 games played). Was the poor performance within the normal range of fluctuations? Answer: Over a span of 10 games, a team that normally plays between 0.500 and 0.600

Table 7.6. Minimum number of wins expected in a given sample
of games played

Number of consecutive games	Normal win percentage				
	0.300	0.400	0.500	0.600	0.700
5	0/1	1/1	1/2	2/2	2/3
10	1/2	2/3	3/4	4/5	5/6
15	2/4	4/5	5/6	7/8	8/10
20	3/5	4/7	5/9	9/11	11/13
30	6/8	9/11	11/14	15/17	18/20
40	8/10	12/14	16/18	20/22	24/27
82	19/22	27/30	35/39	43/47	52/55

Note: Results are given for confidence levels of 90 percent/70 percent.

should reap at least 4 or 5 wins 70 percent of the time and at least 3 or 4 wins 90 percent of the time. The odds of winning 4 or fewer are therefore between 10 and 30 percent. So the odds of winning 4 out of 11 are about 25 percent, not that unlikely.

2. The Montreal Canadiens fired head coach Jacques Martin on December 17, 2011, after the team claimed only 13 wins after 32 games. The Canadiens had a 0.536 win percentage in the previous season, and, since no major change had been made to the team, there were no indications that results would be significantly different in the 2011–12 season. Was the team performance unreasonably low? Answer: Let's take 0.500 as an expected win percentage. After 30 games, there's a 30 percent chance to have 13 wins or fewer and a 10 percent chance to have fewer than 10. After 32 games, the odds of reaping 13 wins are on the order of 25 percent. The poor performance may just have been part of normal fluctuations.

3. The Carolina Hurricanes played for a percentage of 0.488 in the 2010–11 season. After 40 games into the 2011–12 season, they had a dismal 13 wins. Can we conclude definitely that they were a worse team than the year before? Answer: According to table 7.6, there's only a 10 percent chance of having 15 wins or fewer from an expected performance of 0.500. The odds of scoring only 13 are even smaller. It is more likely that the Hurricanes went from being a 0.500 team to being a 0.400 team.

Goalie Production

Like pitchers in baseball, goalies can make coaches and fans nervous. After a small number of shots, a goalie may allow one or two goals more than expected, and sometimes because of no real fault of his own. Sometimes a decision must be made by the coach to replace him. To add to the decisional problem, a goalie doesn't usually ask to be replaced when he doesn't feel well or confident enough; he may actually want stay in and redeem himself. So, like in baseball, it's up to the coaching staff to determine when to replace a goalie (although it happens much less often in hockey than in baseball). To decide, coaches observe the way the goalie reacts to shots, the difficulty level of each shot, and so on. But can the number of goals and the number of shots be used to make a judgment call? For example, if the goalie blocks 20 shots, then lets in 3 of the next 5 shots, do we pull him out? What about 3 goals in the first 5 shots? The likelihood of these events can be calculated with the laws of statistics.

The results shown in figure 7.9 will help answer such questions. Based on the statistics of random events, the table gives the minimum number of consecutive shots needed to score a given number of goals. These are not absolute minimums but the lowest number of shots required to score the goals with the odds of once per season (or 82 games, or 2,500 shots in total). The performances quoted are very poor indeed, but these are also rare events. For example, a 0.900 goalie may let in 4 goals in 7 shots in a row about once every 82 games. Note that the sequence of shots may be counted at any time during a game, and not necessarily from the start. It is statistically possible (over an entire season) that a 0.900 goalie will block, say, 20 shots in a row, then let in the next 3 shots by chance, without an outside reason like loss of focus. On the other hand, 4 goals in 4 shots or 5 goals in 6 shots would be unlikely. But 2 goals in 2 shots will likely happen more than once in a season.

How Important Is Scoring the First Goal?

A professional hockey player once made an interesting observation just moments after his team was eliminated from the NHL playoffs: "We knew the first team to score was going to win. They scored first,

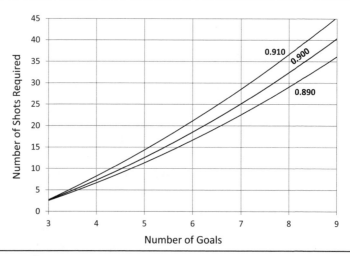

FIGURE 7.9. The minimum number of shots required to score a given number of goals, for save percentages of 0.890, 0.900, and 0.910, with a probability of occurrence of once per season (82 games at 30 shots per game)

and they won." I was surprised to hear that from a top NHL sniper. It was a rather defeatist mindset, I thought at the time.

Let's verify the idea that the first goal scored in a game is more important than other goals scored in regulation time. We start with data taken from the beginning of the 2006–7 NHL season, all teams included, found in table 7.7.

So how important is the first goal? From line 2, we see that the team opening the score in a game wins 70.2 percent of the time (and 75.4 percent of games decided in regulation time). This is a much higher success rate than the 50 percent win average posted by all teams as a whole. From an earlier section, we know that the team that is "supposed" to win (because data favor it) wins between 55 and 60 percent of the time only, depending on how evenly matched teams in the league are. So yes, the first goal has an impact. But since any goal, not just the first one, is helpful toward propelling a team to a victory, it makes sense to ask whether the first goal helps more than other goals.

Let's look at how many goals were scored by winning and losing teams. For all games considered, the winning team scored 65.8 percent of all goals in regulation time (empty netters are not included). In

Table 7.7. Statistics used to determine the importance of scoring first

1. Games considered	248
2. Win percentage when scoring first	70.2
3. Win percentage when trailing first	29.8
4. Games decided in regulation time	195
a. Games won by the team scoring first	147
b. Number of goals scored	1,133
c. Number of shutouts	32
d. Empty netters	30
e. "Regular" goals (that is, not empty netters)	1,103
f. "Regular" goals per game	5.66
i. by the winning team	3.96
ii. by the losing team	1.69
5. Games decided in overtime (shootouts included)	53
a. Games won by the team scoring first	27
b. Total number of goals in regulation time	300
c. Number of shutouts	0
d. Average number of goals in regulation time	5.66
i. by the winning team	2.83
ii. by the losing team	2.83

other words, if one were to pick at random a goal scored in regulation time, that goal will be associated with a win 65.8 percent of the time. But the first goal is associated with a victory 70.2 percent of the time, so we conclude that the first goal is indeed slightly more important. The difference is hardly a reason to call it quits when trailing by 1 goal, however.

We can speculate on a number of reasons why the first goal is slightly more important than other goals. The team that is ahead by 1 goal has the option of folding into a defensive mode, especially toward the end of the game, while the trailing team has to open play and take more chances. Confidence may play a role, as well as the stress of being behind while the clock is ticking.

Of course, it matters when the first goal is scored too. When it happens late in a game, there is less time for the other team to tie the score, and the leading team has a greater advantage. While 70.2 percent can be taken as the average, however, the percentage may be much higher when the first goal is scored in the third period, for example.

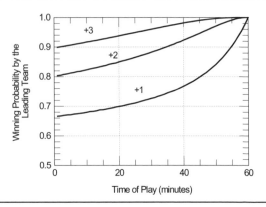

FIGURE 7.10. Winning probabilities by a team when it leads by one, two, and three goals after a given number of minutes. The two teams are assumed to be of equal strength.

We can also examine this problem in more general terms: what are the odds of winning a game when leading by one goal (or more) after N minutes of play? We can verify this with computer simulations, the results of which are plotted in figure 7.10. Here, the simulation assumes both teams usually score at the rate of 1 goal per 22.2 minutes of play (or 2.7 goals per game, on average). The two teams are therefore of equal strength. We see that scoring the first goal at the beginning of play wins the game 66 percent of the time, and when leading by 1 after 2 periods of play (40 minutes), it jumps to 76 percent.

How Often Does a Playoff Series Decide the Stronger Team?

It is generally agreed that a single game does not decide which team is the better one. But when a team wins a playoff, say a best-of-seven series, we can say with more confidence the better team came out on top. But does the better team always win a series? If we agree that a weaker team has a better than zero chance to win any game against a stronger team, we have to admit that it is possible that it can win a second game, and a third, and so on. Although the odds may be small, the weaker team could win any playoff series of any length.

In this section, we estimate the chances of Team A, having a win percentage w_A, of winning a series against Team B with a win percentage w_B. In doing so, we ignore the home ice advantage factor.

The odds of Team A winning over Team B are given by the equation at the beginning of the chapter. From these odds, a series between the two teams can be simulated on a computer by generating random numbers, using something called the "Monte Carlo method" (because of the association of the place with gambling). The process is repeated many times to obtain the proportion of series won by Team A. I ran such simulations of a series of the best of any given number (N) series, with N equaling 1, 3, 5, 7, and 9. (Even though a nine-game series is never played in the NHL, it is included for comparison purposes.) I used only winning percentages between 0.300 and 0.700 to represent typical situations. The results are plotted in figure 7.11. As it turns out, two factors affect the winning chances of the better team (the one with the higher win percentage): (1) the difference in win percentage ($w_A - w_B$) and (2) the series length. As the figure shows, if the teams differ by 20 percent, the better team comes up on top 70 percent of the time in a one-game series and 85 percent of the time in a best-of-seven series. The results for the best-of-seven series agree with a formula derived by others.[1]

The winner of the Stanley Cup needs to survive four consecutive best-of-seven series against teams that are typically close to them in strength. Because of this, the chances for the best overall team's winning the Cup are not that high, probably much less than 50 percent. Indeed, from 1993 to 2013, the top-ranked team won the Cup 6 out of 20 times (by Colorado, Dallas, the New York Rangers, Chicago, and twice by Detroit). If the league wanted to reward the best team, it would make sense to confer the Stanley Cup on the team finishing first overall in the standings at the end of the season. But the point of a playoff is the excitement of a fresh race and of potential upsets.

---

1. If p is the probability the stronger team wins any game (so that $p \geq 1/2$), then the probability of that team winning a best-of-seven series is $(1 - p)^4 (1 + 4p + 10p^2 + 20p^3)$. See P. J. Nahin, *Duelling Idiots and Other Probability Puzzlers* (Princeton, N.J.: Princeton University Press, 2002), 26, 91–94.

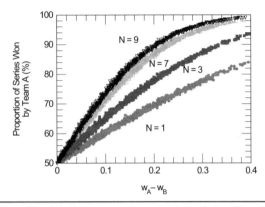

FIGURE 7.11. The odds of the better team winning a best-of playoff improve with the number of games played. Each point is a simulation of 10,000 series between 2 teams with randomly chosen win percentages (but limited to between 0.300 and 0.700).

Let's examine the best-of-seven series in more detail by including home ice advantage. As mentioned previously, hockey teams tend to win more often at home than on the road. In the current best-of-seven NHL playoff format, the team with the higher position in the standings plays at home for games 1, 2, 5, and 7 (though games 5 and 7 may not be necessary). Starting the playoffs at home can give two advantages: a possible early lead and a final and decisive game 5 or 7 played at home. As we have already seen, the winning percentage of NHL teams at home is, on average, higher than their overall win percentage by a factor of 1.12. On the road, it is lower by a factor of 0.88. When two teams of equal strength meet in playoffs, each team would have a 50 percent chance of taking the series if all things were equal, but with home ice, the chances are tilted to 70 percent in favor of the team starting at home. It's not a guaranteed victory, but it helps.

Relax Now, You've Won the Cup

While we're on the topic of winning the Stanley Cup, here's another question: does winning the league championship boost a team's performance in the following season? If winning the Stanley Cup is a morale and confidence booster, it does not show in the statistics of winning, at least not in the immediate aftermath. As table 7.8 demonstrates, the performance of Stanley Cup defending champions tends

Table 7.8. Performance of Stanley Cup winners and runners-up at the start of the following season

Season after winning the cup	Stanley Cup defending champions	Win % in Cup-winning season	Win % in first 5 games of next season	Runners-up	Win % in Cup finals season	Win % in first 5 games of next season
2013–14	Chicago	0.75	0.60	Boston	0.58	0.60
2012–13	Los Angeles	0.49	0.40	New Jersey	0.59	0.60
2011–12	Boston	0.56	0.40	Vancouver	0.66	0.40
2010–11	Chicago	0.63	0.40	Philadelphia	0.50	0.40
2009–10	Pittsburgh	0.55	0.80	Detroit	0.62	0.40
2008–9	Detroit	0.66	0.60	Pittsburgh	0.57	0.40
2007–8	Anaheim	0.59	0.20	Ottawa	0.59	1.00
2006–7	Carolina	0.63	0.20	Edmonton	0.50	0.60
2005–6	Tampa Bay	0.56	0.40	Calgary	0.51	0.20
2003–4	New Jersey	0.56	0.20	Anaheim	0.49	0.00
2002–3	Detroit	0.62	0.60	Carolina	0.43	0.20
2001–2	Colorado	0.63	0.60	New Jersey	0.59	0.20
2000–2001	New Jersey	0.55	0.40	Dallas	0.52	0.40
1999–2000	Dallas	0.62	0.80	Buffalo	0.45	0.00
1998–99	Detroit	0.54	0.80	Washington	0.49	0.40
1997–98	Detroit	0.46	0.80	Philadelphia	0.55	0.60
1996–97	Colorado	0.57	0.40	Florida	0.50	0.60
1995–96	New Jersey	0.46	0.80	Detroit	0.69	0.60
1994–95	Rangers	0.62	0.20	Vancouver	0.49	0.20
	Averages	0.58	0.51		0.54	0.41

to suffer at the beginning of the following season. It's happened to 14 out of the last 19 Stanley Cup champions. Not only did their win percentage after five games into the new season drop, it was also just average or below average compared with the league. Of course it's not a rule, but the trend is there. The problem has also afflicted 14 out of the last 19 Stanley Cup runners-up.

Why don't champions live up to their reputation in the following season? A number of possible explanations can be put forth. One possible reason is fatigue. With an already long regular season stretched farther by almost two more months, rest and vacation time are shortened. But one could imagine that young athletes could fully recover after a couple weeks of rest. Supporting this counterargument is the fact that the performance drops also when the following season is delayed, like during the post-lockout seasons of 1994–95, 2005–6, and 2012–13, when the performance of defending champions and finalists also dropped in five out of six cases.

An alternative explanation is that winning teams "rest on their laurels" a bit too much. To a hockey player, winning the Stanley Cup is a childhood dream, the ultimate career goal. Once the goal is achieved, there might be less motivation to work as hard the following season. (This, of course, until reality sets in and the Cup is up for grabs all over again.) This does not explain why the Stanley Cup runners-up are also impacted negatively, however. We would expect the losing team to come back with a vengeance, hungry for a second chance. Or perhaps in their case, reaching the championship final and losing it has a lasting negative effect on morale.

O Canada, Where's Lord Stanley's Cup?

As a Canadian and a hockey fan, I have to bring up another statistical question that it almost pains me to ask: why hasn't Canada—which has more NHL players than other countries combined and which practically invented the game—won more championships in recent years? It's been a long time since Lord Stanley's trophy spent a summer up north as part of a Canadian winning team. Since the Montreal Canadiens took the championship in 1993, no other Canadian hockey club has won the trophy. Historically, this looks like an anomaly. As of 2014, Canadian teams had won the trophy 41 times out of 87, a success rate higher than the proportion of Canadian hockey teams in the league, which averaged 30 percent over the same period. (For our purposes, we are assuming 1926–27 as the starting season, the year the NHL took over the Stanley Cup.) Before the current drought, the longest winless streak was just six seasons long, and that was back in 1941.

This is hard to take for a nation whose unofficial national sport is hockey. Hockey is to Canadians what baseball is to the Americans and what soccer is to the Brazilians and the English (who, incidentally, have not won a World Cup since 1966). It is not unusual in Canada to sell out an arena during the playoffs even when the team is not playing at home. Fans will cheer watching their hockey heroes on the big screen. Of course, when complaining about not having won the Cup in a long while, Canadians don't mention the 15 times they won it the 1960s and 1970s.

But is there something to this long Canadian drought? Assuming that all teams have the same chance to win the Cup, what would be the odds of Canadian teams not winning in more than two decades?

Let's look at the numbers. From the 1993–94 season until the 2013–14 season, there were 20 seasons played (the 2004–5 season was canceled). All things being equal, the odds for the Cup to have been won by an American team each time are half of 1 percent (or 0.5 percent). In other words, back in 1993, if one were to have made a bet that no Canadian team would win the Stanley Cup over the next 20 seasons, the odds would have been 99.5 percent against this bet.

Sure, it did not help that the proportion of Canadian teams went from 8/26 (31 percent) in 1993 to 6/30 (20 percent) in 2000, but, even taking this lower representation into account, Canadian teams should have made the playoff finals 9 times (the expected average) but they made it only 5 times. They should have won the Cup 4 times but took none. If coincidence is to blame, it's become harder to justify each year that goes by.

From the 1993–94 season until the 2013–14 season, there were 320 playoff berths available, and Canadian teams clinched 67 of them, somewhat below expectation (72 would equal their representative proportion). Once in the playoffs, Canadian teams increasingly under-performed. Out of the Canadian teams that made the playoffs, 43 percent reached the quarterfinals, below the expected 50 percent. And out of the teams that made it to the quarterfinals, a dismal 34 percent made it to the semifinals, again below the 50 percent expectation. And from there, 50 percent made it to the Stanley Cup final and none of them won in 5 attempts.

Part of the reason may be economics: the low value of the Canadian dollar for many years made it harder for Canadian teams to offer lucrative contracts to star hockey players. The salary caps introduced by the NHL have alleviated this problem to some extent by increasing parity in the league. One other possible explanation is the long life of hockey dynasties. The makeup of a hockey team takes several years to change significantly, so the same few strong teams may dominate for a while. An example is the Detroit Red Wings, who have secured a playoff spot every season since 1991 and won the Stanley Cup four times since then.

Furthermore, Canadian markets can be demanding on players and team personnel. In a way, this is normal, as the popularity of a sport comes with the cost of increased pressure and expectations. While it remains to be proved that it affects the performance of Canadian teams adversely, when every game and every decision by coaches is scrutinized, analyzed, and criticized, it would not be surprising if it were so. For that reason alone, I imagine it might be easier living off hockey in places like Tampa Bay than in Montreal or Toronto. On the other hand, if fan pressure were the culprit, it would also seem logical that Canadian teams would benefit less from playing at home in front of their demanding fans. But it's not the case: from 2005 to 2014, both Canadian and American teams have won 30 percent more games at home than on the road.

Canadians take solace in the facts that they still produce the most hockey talents and they win more hockey international tournaments, including the Olympics, than any other country. But one should not overlook this slightly disturbing fact: the vast majority of players on the Canadian National Olympic Teams play for NHL teams in American cities, and many of them have been living and playing in the United States since they were teenagers. In the past two Winter Olympics (2010 and 2014), only 6 of the 48 players were from Canadian NHL teams at the time whereas it should have been more than 10 for equal representation by proportion of Canadian NHL teams.

Yet another explanation could be that when a sport reaches some critical mass of importance in the eyes of the public, the freedom to experiment and innovate is hampered. When every layman is an expert, decisions by team managers are under constant scrutiny, and this may induce fear of risk and promote inertia. A hockey system that is big and important can't adjust swiftly to new realities. Sport-crazed societies also have the tendency to select a few individuals and raise them to the level of authorities—or demigods so to speak—and this reverence by the masses toward a few makes the system even less open to new ideas. To fall back and settle for old-school approaches might be the easiest thing to do in such environment.

Levels of Hockey

It's always interesting to watch teams from different hockey leagues play each other. Not only is it a chance to compare different playing styles and hockey systems, it's also an opportunity to compare the level of the game in each league. Since 2007, NHL teams have played a number of friendly, preseason games against European elite teams, an event called the "NHL Premiere." There are also international tournaments, such as the Winter Olympics and the IIHF World Championships, opposing players from various leagues. Some Olympic teams, like Team Canada and Team USA, are composed exclusively of NHL players, whereas other teams, from countries like Norway, Germany, and Latvia, are made up mostly of European league players. Some teams are a mix of European and Russian leagues and NHL players. So when the American and Norwegian national teams play each other, in a way it is a competitive game opposing the NHL and the European leagues.

One would think that NHL games are the very best and highest levels of ice hockey, but they're not. It is the Winter Olympics that boasts the best talents. Some national teams are actually made up of just NHL stars. Of course, the NHL All-Star Game boasts even better rosters than the best Olympic team, but the competitive edge is not the same. The point of the All-Star Game is to display skills and to have fun. At the Olympics, it's about pride and winning a medal for your home country, and that makes the difference. Nonetheless, when it comes to hockey leagues, the NHL is widely regarded as the most competitive hockey circuit in the world. The talent it attracts is correlated to the size of its market and the salaries it can pay. We can say all we want against the problem of big money in the sport—and certainly hockey fans have felt the sting of being pushed aside while millionaires fight over their money—but so far it's been the most effective way to attract and retain the best hockey talents. The Russian elite league, the Kontinental Hockey League (KHL), would come in at a close second in ranking. It has several players who could play in the NHL—some would be NHL stars indeed—and a few of its teams would compete well in the NHL. Not far behind are some of Europe's top leagues, particularly in Sweden, Finland, Switzerland, and the Czech Republic. They have levels comparable to that of the

American Hockey League (AHL), where NHL farm teams operate, although the European style of play is less physical due in part to the larger size of the rink.

Other than ranking them, can we say more about how hockey leagues compare with each other, in absolute terms? Does the typical KHL player play at 90 percent the level of the typical NHL player, or is it closer to 80 or even 70 percent?

We may try to answer this question by looking at the results of interleague games. During the last NHL Premiere games, out of the 28 friendly games NHL teams have played against European teams, the NHL has won by a ratio of 6 games to 1. Can we conclude, then, that European teams play at one-sixth the ability of NHL teams? Of course not. Game results don't translate so easily into levels of strength for many reasons, not the least of which is the game-to-game randomness and variability (as discussed in previous sections). A score of 8-0 doesn't mean one team has hockey skills and the other one has none.

We have a tendency to overestimate the strength of teams and players that are better than others. (When I was working on this book, a game was being played between Team Canada and Team Norway at the 2014 Winter Olympics in Sochi, Russia. Before the game, sports analysts didn't give much chance to Norway, but they proved to be worthy and determined opponents. Norwegian goalie Lars Haugen of the KHL's Minsk Dynamo, who apparently had not played for two months because of an injury, looked very sharp. Norway seemed to be matching Canada's speed, which in turn seemed to dominate in physical strength and endurance. Canada went on to win 3-1.)

One interesting way to gauge the absolute level of hockey teams in various leagues is to consider statistics from games played in situations of imbalance. When a player receives a penalty, his team plays one man short, the result of which is a diminished chance to score (and an increased chance to be scored on). But this imbalance does not mean a goal is guaranteed for the team with the power play. In the NHL, a power play goal (five players against four) is scored once out of every five minor penalties, on average. Perhaps more surprising, once in a while, there is a goal scored by the shorthanded team—a great morale booster for them and a sure occasion to celebrate. But in the long run, the odds strongly favor the team with the extra player.

Let's make this assumption: let's assume that taking one player out reduces a team's strength to four-fifths (80 percent) of its original strength. In other words, taking one player out diminishes the team's offensive ability by 20 percent. This of course assumes that a team is just the sum of its parts, and, while this is not strictly true, we'll take it as a starting point.

We can now combine this estimate with the statistics of playing shorthanded.[2] When NHL teams play 5-on-5, each team produces a goal every 27 minutes, on average. So, if the whole game were played 5-on-5 for 60 minutes, the final score would average 2.3 goals apiece. (In reality an average of 2.7 goals is scored because of penalties.) But in a shorthanded situation (4-on-5), a team scores a goal every 74 minutes, on average, while the team on the power play scores every 9.7 minutes. So a 20 percent mismatch makes a large difference in goal production. To be specific, if the entire game were played 5 against 4, a typical score would be 6-1 (and sometimes much worse). The odds of the shorthanded team's winning, according to the laws of statistics, would be only 1 in 100.

As we will see, a 20 percent difference in strength is actually quite large, at least larger than the typical difference between teams at the highest levels of hockey. To examine situations where teams differ by less than 20 percent, we use a mathematical trick called a "linear interpolation." The result is shown in figure 7.12. The horizontal axis is the relative difference in team strength expressed decimally (for example, 0.05 = 5 percent difference), while the vertical axis shows the predicted percentage of times the stronger team wins. As expected, on even strength (a strength difference of zero), the result is a 0.500 win percentage. But the balance tips very quickly: with a 10 percent strength difference, the better team wins 90 percent of the games. If we use this graph to estimate the levels of the NHL and European teams based on the 28 games played in recent years (86 percent win percentage by the NHL), the difference in their levels is 8 percent.

We can also look at the problem from the point of view of the goal differential in a game. As mentioned earlier, a 20 percent difference in strength leads to a score difference of about five goals. Using this

2. D. Beaudoin and T. Swartz, "Strategies for Pulling the Goalie in Hockey," *The American Statistician* 64, no. 3 (2010).

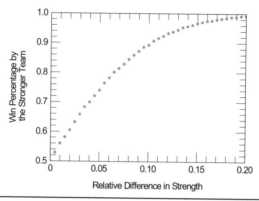

FIGURE 7.12. Win probability by a stronger team over a weaker one as a function of relative strength difference (for example, 0.15 = 15 percent difference)

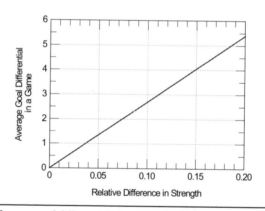

FIGURE 7.13. Average goal differential in a game played between teams with a given strength difference

as reference, we can project the goal differential (per game) when teams are uneven. Figure 7.13 shows the results. It is interesting to note that an 8 percent difference in team strength results in a 2-goal deficit by the weaker team. As it turns out, the average score of the 28 NHL Premiere games played so far was 4.5 goals to 2.2 goals in favor of NHL teams. This supports the assumption we made at the beginning.

In chapter 4 we estimated the change in fitness among hockey players as they age. Figure 4.8 suggested that players lose their ability to play rather slowly, and they are still within 10 percent of their peak abilities when they reach 40. Given that the top professional leagues are also within 10 percent in strength, it is not surprising to see some older NHLers joining other top leagues and finding the level of play to be suitable for them.

The relatively small—and some would say surprisingly small— difference in player and team strengths between the NHL and European leagues confirms a phenomenon well-known to sports scientists, namely that absolute differences in skills tend to be small at the highest levels of a sport. And the differences tend to get smaller as the sport grows in popularity and the population of players increases. It is always easier to stand out and dominate as a player when fewer people play a sport. As explained by Stephen Jay Gould in the case of the disappearance of 0.400 hitting in baseball,[3] variability and extreme values of performances tend to diminish with an increasing size of the baseball-playing population. This is yet another version of the law of small numbers.

This concludes the book. I hope I have given you a taste of how complex and wonderful the game of hockey can be, and with it, some tools to perhaps understand the science of the game a little better. Whether you're a fan or a player looking to improve his or her game, I hope you found in it something useful or entertaining.

Even though I have covered many aspects of the game, there is much more to be discovered. Each year hockey players find new tricks and new ways to dazzle us, and sports scientists and equipment companies keep improving the technical aspects of the game. Coaches and hockey thinkers invent new strategies to find every possible edge. And with the internationalization of hockey, this trend will surely continue. Hockey rinks are being built around the world at a record pace, in countries where, not so long ago, ice hockey was unknown to most. As international tournament organizers can attest, the number of hockey players is growing and lifting the level of hockey played around the world. This can only spell good things for the future of the game.

---

3. S. J. Gould, *Full House: The Spread of Excellence from Plato to Darwin* (New York: Harmony Books, 1996).

Converting Metric to Standard Units

To obtain	Multiply	By
mph	km/h	0.621
km/h	mph	1.61
m/s	mph	0.447
m/s	km/h	0.278
m	ft	0.305
ft	m	3.28
in	cm	0.394
cm	in	2.54
U.S. gallon	liter	0.264
liter	U.S. gallon	3.79
lbs	kg	2.2
kg	lbs	0.453

Symbols and Acronyms Used

ACL   anterior cruciate ligament

AHL   American Hockey League

ATP   adenosine triphosphate

g acceleration due to gravitational pull

GAA   goals against average (average number of goals scored against a goalie in a game)

Hz   Hertz, a measure of frequency, the number of vibration cycles per second

IIHF   International Ice Hockey Federation

KHL   Kontinental Hockey League

N   Newton, a standard unit of force

NCAA   National Collegiate Athletic Association

NHL   National Hockey League

OHL   Ontario Hockey League

PPG   points per game by a player

SHL   Swedish Hockey League

USHL   United States Hockey League

W   watt

The Probability of Winning

Here you will find an explanation of the origins of equation (1) found in chapter 7. To derive this equation, we make two basic assumptions:

1. The winning percentage of Teams A and B, given by w_A and w_B, are measurements that are proportional to each team's strength. This approximation should be valid when both teams play against the same pool of teams in a given league.
2. Team A's chances of winning against Team B is proportional to Team B's losing percentage $1 - w_B$, and vice versa. This could be called the "window of opportunity" offered by Team B to Team A. Of course, if Team B never loses, $1 - w_B = 0$ and Team A doesn't have a chance.

According to these assumptions, Team A's chances of winning are proportional to both w_A (its own strength) and to $1 - w_B$ (its window of opportunity). So the quantity $A = w_A(1 - w_B)$ should be proportional to the winning chances of Team A, and for Team B it is $B = w_B(1 - w_A)$.

The final step is to find p (the winning probability of Team A) by making a direct comparison between quantities A and B. In statistical theory, the probability of an event occurring is the ratio between the number of successful events and the number of all possible events. In this case, p must be the ratio between A, which scales as the number of times A will win, and $A + B$, which scales as the sum of all possible outcomes (A or B winning). As a result:

$$p = \frac{A}{A+B} = \frac{w_A(1 - w_B)}{w_A(1 - w_B) + w_B(1 - w_A)}$$

Note that this equation assumes w_A and w_B are given in decimal formats (with values between 0 and 1), as is the case with p.

INDEX

Page numbers in *italics* refer to figures and tables.